Fundamentals of Polymer–Clay Nanocomposites

Written for graduate students, researchers, and practitioners, *Fundamentals of Polymer–Clay Nanocomposites* provides a complete introduction to the science, engineering, and commercial applications of this new class of material.

Starting with a discussion of general concepts, the authors define in detail the specific terms used in the field, providing newcomers with a strong foundation in the area. Then, the physical and mechanical properties of polymer–clay nanocomposites are described, with chapters on thermodynamics and kinetics, engineering properties, barrier properties, and flame retardancy. Mechanisms underpinning observed effects, such as UV resistance, solvent resistance, and hardness, are also explained. Throughout each chapter, experimental results are combined with theory, ensuring that the reader gains a full appreciation of the subject matter.

In-depth discussions of clay and clay surface treatment, fabrication, and characterization of nanocomposites are provided, and particular emphasis is placed on the proper use and interpretation of analytical techniques, helping the reader to avoid artifacts in their own work. With commercial applications discussed throughout, this is an ideal reference for those working in polymer science.

Gary W. Beall is a Professor in the Department of Chemistry and Biochemistry at Texas State University, San Marcos, and Director of the Center for Nanophase Research. He was awarded his Ph.D. in Physical Chemistry from Baylor University in 1975, and has since gained over 20 years of industry experience, co-edited the first book on polymer–clay nanocomposites, and authored numerous technical papers and two book chapters.

Clois E. Powell is Associate Director of the Center for Nanophase Research at Texas State University, San Marcos, where he has worked since 2004. After receiving his Ph.D. in Organic Chemistry from Rutgers University in 1979, he gained over 20 years of professional experience with companies including Southern Clay Products Inc. and The Sherwin Williams Company.

Fundamentals of Polymer–Clay Nanocomposites

GARY W. BEALL AND

CLOIS E. POWELL

Texas State University, San Marcos

CAMBRIDGE
UNIVERSITY PRESS

CAMBRIDGE UNIVERSITY PRESS
Cambridge, New York, Melbourne, Madrid, Cape Town,
Singapore, São Paulo, Delhi, Tokyo, Mexico City

Cambridge University Press
The Edinburgh Building, Cambridge CB2 8RU, UK

Published in the United States of America by Cambridge University Press, New York

www.cambridge.org
Information on this title: www.cambridge.org/9780521876438

First published 2011

Printed in the United Kingdom at the University Press, Cambridge

A catalog record for this publication is available from the British Library

Library of Congress Cataloging-in-Publication Data

Beall, G. W. (Gary W.), author.
Fundamentals of Polymer-Clay Nanocomposites / Gary W. Beall and Clois E. Powell.
p. cm
ISBN 978-0-521-87643-8 (Hardback)
1. Nanocomposites (Materials) 2. Polymer clay. I. Powell, Clois E. (Clois Elbert), 1944–, author. II. Title.
TA418.9.C6B43 2011
620.1′92–dc22

2010044487

ISBN 978-0-521-87643-8 Hardback

Contents

1 Introduction

Can one imagine the utility of a dispersed-phase reinforcement for polymers that has a thickness of 1 nm, a platelike morphology with minimal dimensions of 150 to 200 nm, robust with a modulus of 180 GPa, nontoxic (FDA classification of GRAS; generally regarded as safe for a majority of applications), a surface area in excess of 750 m^2/g, a charge suitable for altering its hydrophobic–hydrophilic balance at will, and a refractive index similar to polymer so that the nanoparticles will appear transparent in the polymer composite? How difficult would it be to prepare such a particle?

This particle is naturally occurring and found around the world. It is easily mined and purified. The reactor for the particle was a volcano. The ash from many volcanoes was spread around the earth during an intense period of activity many millions of years ago. This ash was transformed into clay (montmorillonoids or smectites) by natural processes, into uncharged species (talc and pyrophyllite) and charged species through isomorphic substitution of the crystal structure (hectorite, montmorillonite, saponite, suconite, volchonskoite, vermiculite, and nontronite).

Montmorillonite serves as the principle mineral for the development of polymer–clay nanocomposites discussed in this book. A misunderstanding of the terms bentonite (the ore or rock) and montmorillonite (the mineral) are pervasive in the literature. We will focus on utilizing the mineral name. The composition of montmorillonite can be described by imagining a sandwich structure with the top and bottom layers composed of silica dioxide tetrahedral structures. The center layer is composed of a metal oxide octahedral layer. Metal in this octahedral structure originally was completely Al^{+3} and uncharged (pyrophilite). Isomorphic substitution of Mg^{+2} for Al^{+3} in the octahedral layer produced montmorillonite.

Notice that a substitution of a +2 Mg for a +3 Al will produce a net negative charge. The amount of substitution will dictate the charge on the particle. Nature will not abide a net charge development without supplying counterions for that charge. These are usually sodium, calcium, potassium, etc. ions where the montmorillonite is natural.

Hectorite is the isomorphic substitution of Li for Mg (from talc). Some isomorphic substitution is observed in the tetrahedral silica dioxide layer with Al substituting for Si. These particles are found in nature in stacked arrays similar to a pad of paper. They can be easily mined and purified by dispersing the montmorillonite into water and removing the larger, heavier particles (sand, gravel, etc.) by centrifugation.

Individual montmorillonite particles in water must be obtained by a high shear environment. The nature of the counterions can interfere with ultimate separation of single particles (full exfoliation).

Sodium has the greatest utility as a counterion for the preparation of fully exfoliated montmorillonite dispersions. Sodium can be exchanged for Mg and Ca counterions by standard ion exchange methods (for example, water-softening equipment). These fully exfoliated montmorillonite particles can be exchanged with positively charged organic molecules (for example, quaternary ammonium ions) to form very stable individually organically modified montmorillonite particles. When the modification is significantly hydrophobic, the organomontmorillonite will separate from the water. This material can be filtered from the water, dried, and ground to a powder.

Different organomontmorillonites are employed extensively for a wide range of market areas. An international awareness that organomontmorillonite could be successfully utilized as a dispersed-phase reinforcement in polymers began in the early 1990s.

Many descriptive terms that describe the morphology of montmorillonite in polymers have been misused. The galleries of the organomontmorillonite are composed of the organic treatment separating the individual montmorillonite particles. These galleries begin to fill with polymer and swell. The particle separation up to about 7 nm will allow polymer chains to interact with two particles simultaneously. These structures are referred to as intercalated. Assemblies of these particles are referred to as tactoids and behave as a collection of particles in polymer composites. Separation of montmorillonite particles past this threshold precludes the interaction of polymer chains interacting with more than one particle. These assemblies are referred to as exfoliated polymer–montmorillonite nanocomposites. At this stage, the full benefit of dispersed-phase reinforcement by montmorillonite in the polymer is realized. Because the particles are anisotropic, they also must be aligned to achieve this benefit.

This book will assess the structure–property benefits of these nanoparticles in polymers as a function of their degree of exfoliation, alignment, and dispersion. Methods for the preparation of these polymer nanocomposites will be assessed in relation to the attainment of the optimal morphology for nanoparticle structure within the polymer. This book will not attempt to give an exhaustive review of the literature since many recent works have served that purpose quite adequately. Rather, this book will focus on assessing what is truly understood about polymer–clay nanocomposites and which areas require further study.

2 Thermodynamics and kinetics of polymer–clay nanocomposites

In order for nanocomposites to be useful, they must be thermodynamically stable. It is therefore critical to ensure that clay nanoparticles have surfaces that interact with polymer in a way that yields exfoliated structures that do not spontaneously phase separate. Although some intercalated–exfoliated systems may yield useful improvements in properties, the exfoliated state is still the ultimate goal in producing a nanocomposite with the ultimate property enhancements.

The rate at which intercalation/exfoliation occurs is also of some importance in ensuring that a nanocomposite can be made on a timescale that is commercially viable. Since the level of exfoliation is critical in order that the maximum change in properties in nanocomposites is reached, the ability to measure the level of exfoliation is of paramount importance.

In this chapter, the thermodynamics of intercalation/exfoliation will be discussed in detail, including surface modification of clays, processing strategies, and the enthalpic and entropic components of the intercalation/exfoliation process. In addition, the kinetics related to intercalation/exfoliation will be presented. Finally, a critical evaluation of the analytical methods utilized commonly to determine the level of intercalation/exfoliation will be given.

2.1 Clay surface compatibility with polymers

2.1.1 Smectite clay structure

The discussion of clay surface compatibility with polymers in this section will focus primarily on montmorillonite as the example clay. The characteristics discussed will only vary by degree for other smectic clays.

It is important to understand the structure and general properties of smectic clays. The primary smectite phyllosilicates chosen are analogs

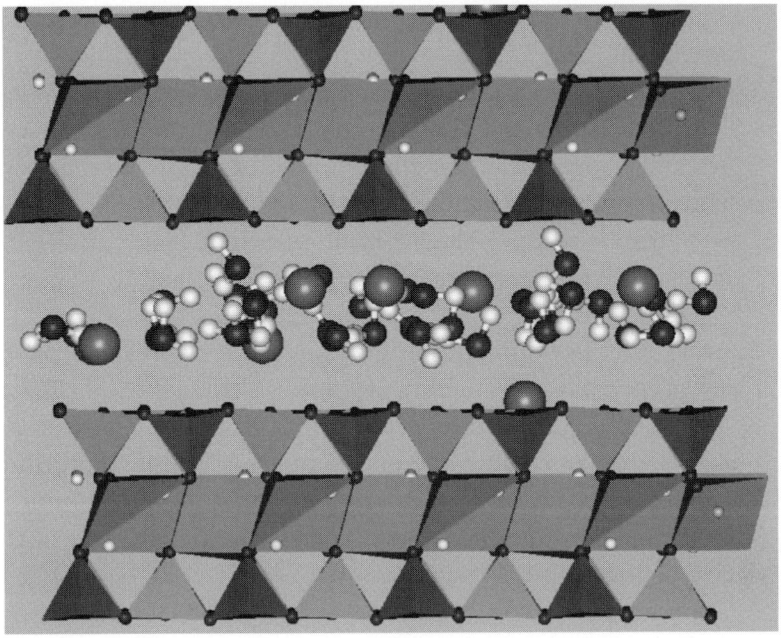

Figure 2.1 Molecular model of montmorillonite showing the octahedral layer of cations sandwiched between two layers of tetrahedrally coordinated silicons.

of talc and pyrophyllite. In pyrophyllite, two silicate layers are sandwiched around a central gibbsite layer. Gibbsite, $Al(OH)_3$, is composed of octahedrally coordinated aluminums surrounded by six hydroxides. The gibbsite layer (G) in pyrophyllite is identical to the structure of gibbsite, except that four of the hydroxides are replaced by four oxygens from the silicate layers (S). The overall structure of pyrophyllite can therefore be imagined as stacked S–G–S sandwiches. Pyrophyllite is identical in physical properties to talc. The two are isomorphous, meaning that they share the same monoclinic structure but have different chemistries. Talc has magnesiums instead of aluminums and is basically indistinguishable from pyrophyllite without a chemical test for aluminum.

Smectic type clays are characterized by a 2:1 structure depicted for the mineral montmorillonite in Figure 2.1. The structure has a central layer of octahedrally coordinated metal ions sandwiched between two layers of tetrahedrally coordinated silicons. This structural motif is shared by a series of minerals that includes talc, pyrophyllite, montmorillonite, nontronite, hectorite, vermiculite, saponite, Beidellite, and the various forms of mica. The differences between these minerals are mainly in structural filling, type of cation, and isomorphous substitution in the octahedral or tetrahedral layer.

Two types of crystallographic filling of the octahedral metal site characterize the structural filling. In the case of talc, hectorite, some vermiculite, and some micas, all three crystallographic sites are occupied by divalent ions such as magnesium. These are normally designated by the term "trioctahedral." The other minerals only have cations at two out of the three sites and are referred to as "dioctahedral." The predominate cation in these cases is Al^{3+} or, in some cases, Fe^{3+}.

The other key difference is the level and type of isomorphous substitution in each of these minerals. Among the trioctahedral clays, talc has no isomorphous substitution, hectorite has moderate substitution of Li^{+} for Mg^{2+}, and vermiculite and mica have high levels of substitution.

In the dioctahedral series, pyrophyllite has no substitutions, montmorillonite has medium amounts of substitution, and vermiculites and mica have high amounts of substitution.

In Figure 2.1, the gray octahedral represent Al^{3+} and the Mg^{2+} ions are randomly substituted in this octahedral layer. These materials have a plate thickness for the 2:1 structure of approximately 1 nm. In the case of talc and pyrophyllite, these 2:1 layer structures manifest themselves in very strong cleavage in one direction and very low hardness. In the case of all other members of the phyllosilicates, the isomorphous substitutions in the lattice result in a charge imbalance that is satisfied by exchangeable cations on the surface. In Figure 2.1 these cations are represented by the sodium ions depicted as space filling. The most common cations in these materials in nature are sodium, magnesium, iron, and calcium. Many of these exchangeable cations are hydrated with water as depicted in Figure 2.1. In the clay minerals, this results in a layered structure that has what are referred to as galleries between plates that contain the exchangeable cation and waters of hydration. Figure 2.2 is a molecular model of this gallery between two plates.

2.1.2 Turbostratic nature of smectite clays

One of the far-reaching effects of this gallery space is that the clay structure is turbostratic with no crystallographic relationship between plates. This is strongly exhibited in the X-ray diffraction pattern where only 00L or HK0 reflections appear with no HKL reflections. A typical X-ray diffraction pattern for sodium montmorillonite is given in Figure 2.3. It can be seen that the broad peak at 7.5° two theta marked A is the peak for the 001

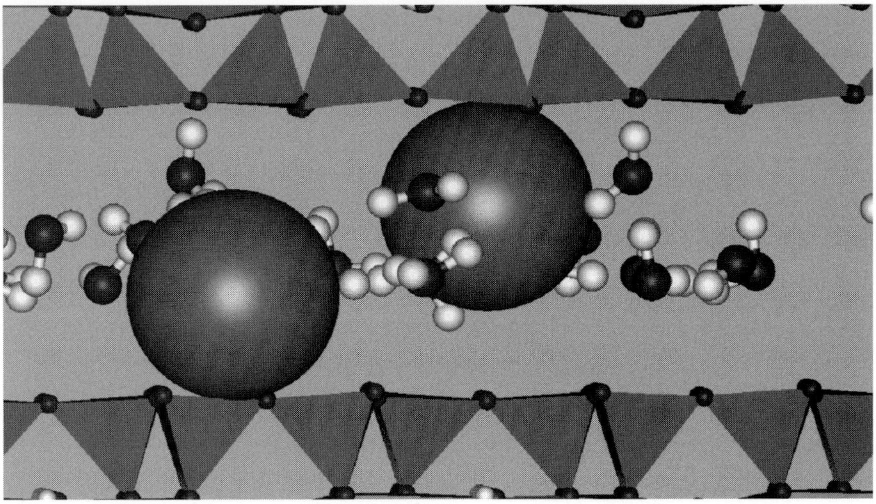

Figure 2.2 Molecular model of the gallery of montmorillonite showing the exchangeable Na^+ cations (displayed as space filling) with their waters of hydration.

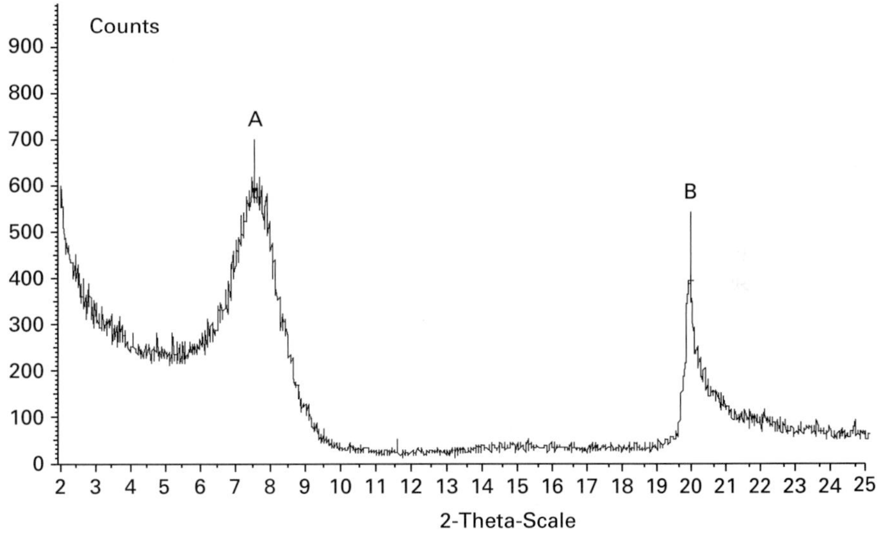

Figure 2.3 X-ray diffraction pattern of Na montmorillonite.

basal spacing which arises from the distance between individual plates, causing it to be variable and the peak at 20° two theta is the 110 Bragg reflection marked B is constant, since it arises from the internal structure of the plates themselves. The fact that the only forces holding the plates together are Van der Waals and hydrogen bonding allows the gallery to expand or contract, depending on what cation is exchanged on the

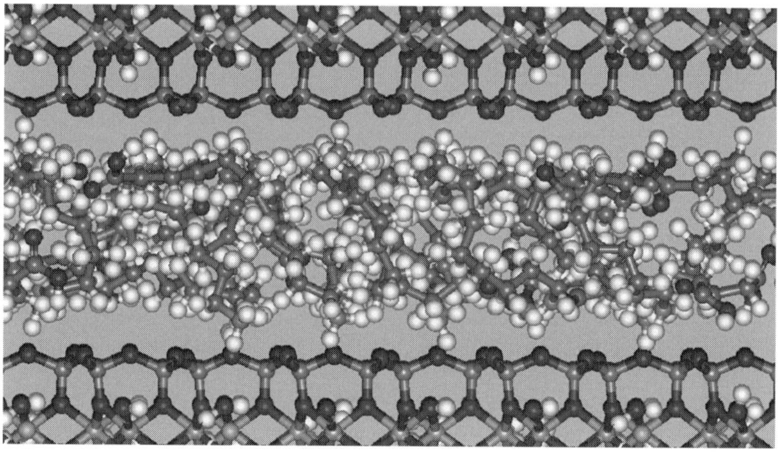

Figure 2.4 Molecular model showing the typical hydrophobic interdigitated structure formed in the gallery when onium ions are exchanged for the metal cation on the clay surface.

surface and depending on the level of hydration. In the extreme case these clay minerals will disperse completely in water-forming thixotropic gels.

2.1.3 Intercalation chemistry

A great variety of molecules can also be substituted for water in the gallery via ion–dipole bonding. Commercially, it is quite common to convert the hydrophilic montmorillonite clay to a hydrophobic state by ion exchanging the sodium for an organic onium ion that contains long alkyl chains. When this is done, the gallery expands and the alkyl chains form an interdigitated structure pictured in Figure 2.4. As mentioned previously, the mineral montmorillonite will be used as the standard example of the family of smectic clays. This is due to the fact that the vast majority of commercial deposits of smectites in the world are montmorillonite. The most common form is the calcium exchanged form but vast deposits of predominately sodium form exist, particularly in the western USA.

One of the biggest mistakes made in conceptualizing polymer nanocomposites is in the relative sizes of the polymer and clay plates. It is common to see schematic representations of wavy polymer molecules that are much longer than the clay plates. This picture is quite misleading and can lead to incorrect interpretation of data. In fact, the normal commercially available montmorillonite has a long dimension in the range of 150 to 200 nm. As an example, if you straightened out a 100000 molecular weight polyvinylpyrrolidone (PVP) polymer, it would be approximately 120 nm long. It is well known that polymers very seldom configure linearly and therefore, the actual size of a random coil is much smaller in most polymers than that of the clay. The only place that the type of picture normally depicted in the literature could apply is with polymers which have molecular weights in excess of 200000, and even in these cases, a random coil would render them the same order of size as the clay plate.

2.1.4 Intercalation of water-soluble polymers

The sodium form of the clay is naturally quite hydrophilic and is not compatible with most polymers. There are a few exceptions to this for water-soluble polymers. It is instructive to look at a few of these polymers and at how they behave with respect to intercalation and exfoliation. It appears that in all of them, the interaction between the polymer and the clay is mainly through ion–dipole bonding to the exchangeable cation on the surface of the clay. This can be seen in the first water-soluble polymer to be discussed, PVP. This polymer has a bulky side group composed of a five-membered ring amide. The carbonyl oxygen bonds to the cation on the surface of the clay, as shown in Figure 2.5. One of the ion–dipole bonds is highlighted at 0.217 nm. It should be noted that the bonded pyrrolidone groups force the pyrrolidone rings into a configuration that makes bonding to the sodium on an adjacent clay plate difficult.

The intercalation behavior of PVP is quite unique and is intropically driven. Intercalation can be obtained simply by first placing powdered sodium montmorillonite in a bowl-type mixer and hydrating the clay with approximately 30 wt.% water. At this stage the clay has the consistency of moist sand. Next, 35 wt.% powdered PVP is then dumped into the mixer and mixed. At first the mixture just appears to be a powdered mess, but within 10 to 15 s of mixing, the whole mass turns into a dough consistency. The dough has formed because the PVP has intercalated into the clay gallery and ion–dipole bonded to the sodium cation on the clay surface. In this process the PVP has displaced many moles of water that were hydrating the sodium. The entropy of the system has increased dramatically, since for every mole of PVP that intercalates multiple moles of water are converted from a bonded state coordinated to the sodium into free liquid water. The intercalation reaction occurs much faster if the

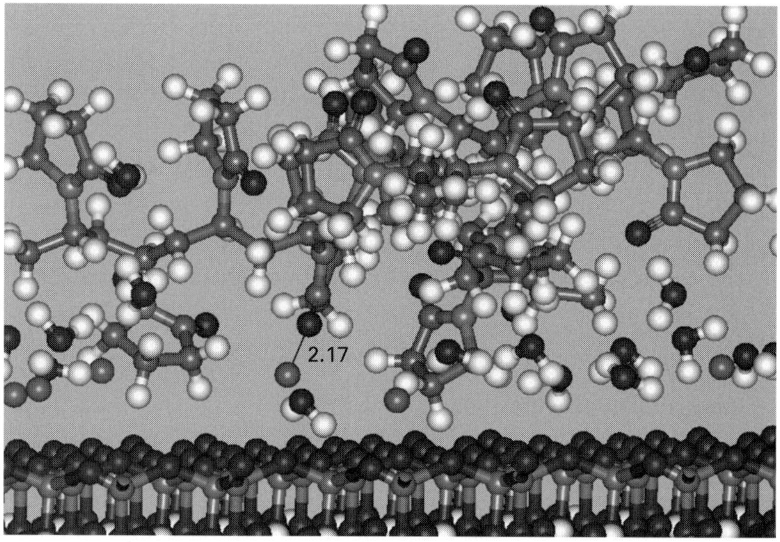

Figure 2.5 Molecular model of polyvinylpyrrolidone bonded via ion–dipole bond to the sodium cation on the surface of the clay.

PVP is dissolved in water prior to mixing it with the clay. The intercalation can be measured easily using X-ray diffraction by following the change in basal *d*-spacing.

It can be seen in Figure 2.6 that there appears to be a stepwise intercalation reaction occurring as more PVP is added relative to the weight of the clay [1]. Molecular modeling indicates that the diameter of the PVP polymer is in the range of 0.93 to 0.96 nm. If we examine the steps in Figure 2.6, the first step goes from 1.26 to 2.4 nm, a change of 1.14. This is not far off the calculated value. The second step has a delta of 0.8 nm and the third, 0.9 nm. The average for all three steps is 0.95 nm, which is in excellent agreement with the molecular modeling.

It appears that the first step corresponds to a monolayer of PVP trapped between the plates of clay. The second step would be a monolayer covering both faces of the plates and which results in a bilayer complex in the gallery between the plates, and the third appears to form a trilayer structure in the gallery. If PVP is added to higher and higher concentrations, the intercalation peak continues to grow and broaden with the eventual complete exfoliation of the clay into the PVP polymer.

The ability to form a stepwise intercalation that leads to exfoliation is unique to PVP. The other water-soluble polymers behave in a different way than PVP that is illustrative of the forces important in producing an exfoliated nanocomposite.

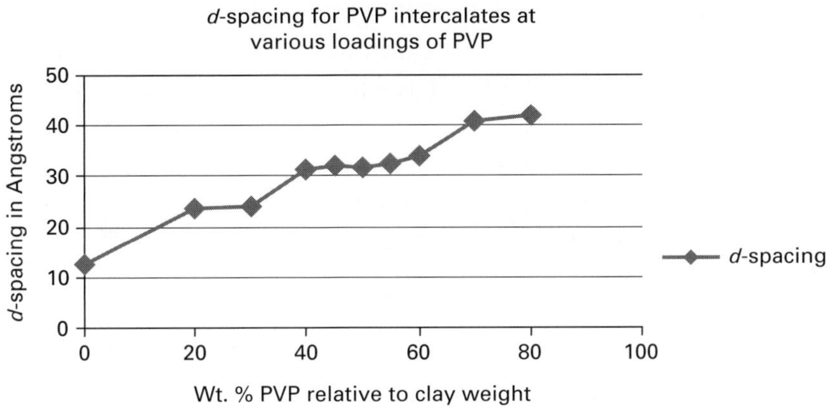

Figure 2.6 *d*-spacing increase caused by the intercalation of polyvinylpyrrolidone into montmorillonite.

A comparative study of the intercalation behavior of PVP, polyvinyl-alcohol (PVOH), and polyethylene glycol (PEG) was conducted. In this study, the type of intercalate formed was monitored by X-ray diffraction. The clay utilized was a high-charge montmorillonite having a cation exchange capacity of 143 meq/100 g. In each case, the polymers were mixed with the clay at 60 wt.% relative to the clay weight, which should ensure the formation of the bilayer structure as seen in PVP. This, however, was not the result obtained. The PVP formed the bilayer structure but the PVOH and PEG both formed only a monolayer. As mentioned previously in Figure 2.5, the PVP, once bonded to the clay surface, forces the other pyrrolidone rings into positions that make bonding to an adjacent plate difficult. This is not the case with PVOH and PEG. The PVOH side groups are not as bulky as the pyrrolidone, allowing more flexibility in the backbone, and the hydroxyl groups are not geometrically as constrained as the carbonyl is in pyrrolidone. This allows the PVOH to bond to both sides of the gallery and essentially to glue the plates together. The same argument applies to PEG, except that the ether linkage makes it even more flexible than PVOH, allowing the chain to configure itself in ways that allow for bonding to adjacent clay surfaces.

It was hypothesized that if the number of cations available for ion–dipole bonding was reduced on the clay surface, the bilayer intercalate could be formed. The clay was exchanged with ammonium ions to replace half of the sodium cations. When the intercalation experiments were repeated, the PVOH now formed the bilayer structure, but the PEG still formed only the monolayer. This behavior for PEG has also been observed in fluoromicas [2]. This would indicate that one important factor in exfoliation is that the polymer would not be able to bond easily to adjacent clay layers. The intercalation/exfoliation characteristics of these water-soluble polymers are interesting; however, for nanocomposites to have widespread usefulness, a broader spectrum of polymers must be amenable to nanocompositing.

2.1.5 Hydrophobic intercalation

In order for the clay to intercalate and exfoliate into a broad array of polymers, the normally hydrophilic surface needs to be modified to a hydrophobic surface that more closely matches the polymers' hydrophobicity.

2.1.6 Intercalation via ion exchange

For montmorillonite, this has been done commercially for decades by ion exchanging the metal cation for organic onium ions. These onium ions are, typically, quaternary ammonium compounds [3,4], but some work has also been reported on phosphoniums [5]. These quaternary ammonium compounds were original commercialized as fabric softeners and antistatic agents. They are typically characterized by at least one long, chained, hydrogenated alkyl derived from either an animal or vegetable source. They therefore have a distribution of chain lengths that is characteristic of the source from which they are derived. Common examples would be hydrogenated beef tallow, which has a distribution of chain lengths centered around C16 and C18 with minor amounts of chains less than 16 or greater than 18, and coconut oil, which has a broad distribution centered around C12 with substantial amounts of chains having 8, 10, 14, 16, and 18 carbons.

The most common quaternary ammonium compounds utilized today to produce organoclays include dimethyl dihydrogenated-tallow, dimethyl benzyl hydrogenated-tallow, dimethyl hydrogenated-tallow (2-ethylhexyl), and methyl tallow bis2hydroxyethyl. The organoclays made utilizing these quats are supplied in some cases at various levels of quat treatment in order to yield a range of hydrophobicity. Some organoclays have been produced utilizing the hydrochloride salts of the analogous amines and, in the case of nylon 6 nanocomposite, the clay is treated with the acid salt of 12 amino lauric acid.

From the general alkyl chemistry shared by all of these quats, except the case of nylon 6, the range of hydrophobicity and surface chemistry is rather limited. The quaternary ammoniums are also somewhat susceptible to thermal degradation and are not approved for direct food contact. These factors have prompted work to develop alternative surface modification chemistries that address some of these issues.

2.1.7 Alternative intercalation chemistries

Alternative clay surface modification schemes can be divided into three main categories: ion–dipole bonding to the cation on the surface of the clay, polymer sorption on the siloxane surface, and edge treatment with silane coupling agents. These different strategies can be combined and, also, augment quaternary ammonium treatment strategies.

2.1.8 Intercalation via ion–dipole bonding

In the ion–dipole approach of surface treatment, advantage is taken of the partial positive charge remaining on the exchangeable cation on the clay surface. This partial charge arises from the fact that the source of the anionic charge of the clay arises in montmorillonite principally from isomorphous substitution of divalent magnesium for trivalent aluminum. This charge deficiency results in an excess of anionic charge in the oxygen layer shared between the tetrahedral silicon layer and the octahedral aluminum layer. This layer of oxygens is approximately 0.6 nm away from the cation on the surface. The charge is therefore quite diffuse at the surface and therefore, it is difficult for the cation to fully satisfy its charge. This is the reason why montmorillonite has such a large hydration energy.

As demonstrated in earlier discussions, water-soluble polymers can ion–dipole bond to these cations and form very stable intercalated structures. This can also be demonstrated with small molecules. The classic example is the displacement of water by ethylene glycol. In this case, the water form of the clay gives a *d*-spacing of 1.26 nm but the ethylene glycol (EG) gives a spacing of over 1.7 nm. The EG intercalate is of little use in polymer nanocomposite because it is too hydrophilic and tends to form a bridging conformation that tends to bond the plates together.

There is a host of other organic molecules that possess both a hydrophilic and hydrophobic portion in the molecule. This is exactly the criteria needed to form an ion–dipole bond to the cation with the hydrophilic end of the molecule forcing the hydrophobic portion away from the surface. A classic example of this can be seen with alkyl pyrrolidones.

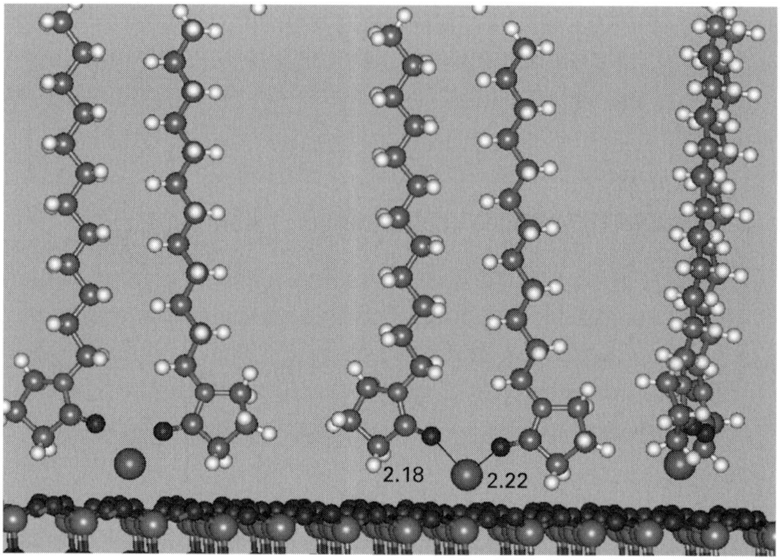

Figure 2.7 Molecular model of dodecyl pyrrolidone ion–dipole bonded to a sodium cation on the clay surface and self-assembling into a rigid post.

Intercalates form readily between the *n*-alkyl pyrrolidones and yield interesting effects as a function of alkyl length and ratio of pyrrolidone to surface cation [6]. When utilizing *n*-dodecyl pyrrolidone (DDP) as a surface treatment, it was found that when DDP was mixed with the clay in a 1:1 ratio to the exchangeable cation, there was an interdigitated intercalate with a *d*-spacing of 2.1 nm, which is quite similar to that observed for traditional quaternary amine treated clays.

The situation changes radically when the ratio of DDP to cation is increased. It was found that when the ratio reaches 2:1, a self-assembly occurs where the two pyrrolidone head groups ion–dipole bond to the cation and then the alkyl chains self-assemble into a rigid post (Figure 2.7). These posts are randomly distributed on the surface so that when plates come together face to face, the posts do not interdigitate.

This can be observed in an increase of the *d*-spacing to 3.5 nm. The self-assembly was observed in alkyl chains as small as eight carbons and strongly reinforced by longer chain lengths of C18.

These observations led to further work to determine if there were compounds that exist on the FDA generally recognized as safe (GRAS) list to facilitate nanocomposites that can be utilized in direct contact with food [7]. In this work, the mono and distearate esters of pentaerythritol (PES) were found for very stable intercalated complexes with montmorillonite that exhibit *d*-spacings of greater than 5 nm. The bonding of the

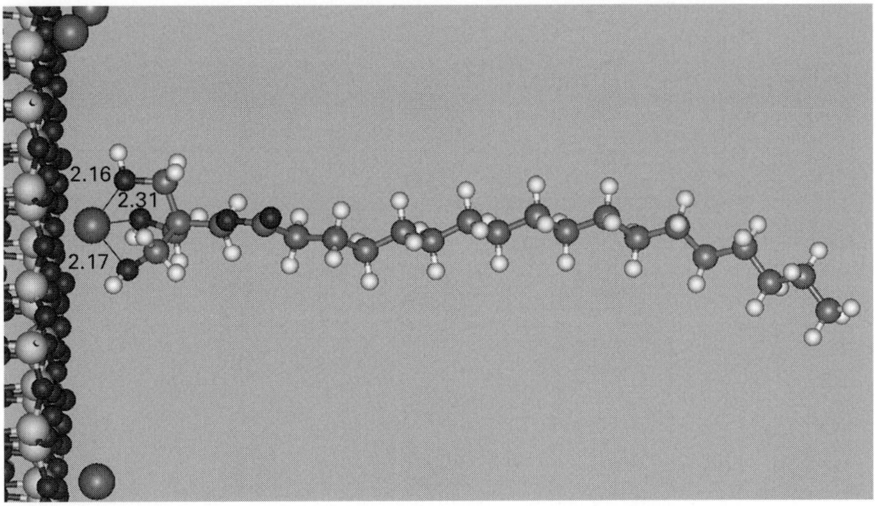

Figure 2.8 Molecular model of a molecule of pentaerythritol monostearate bonded to sodium ion on the clay surface through the three hydroxides.

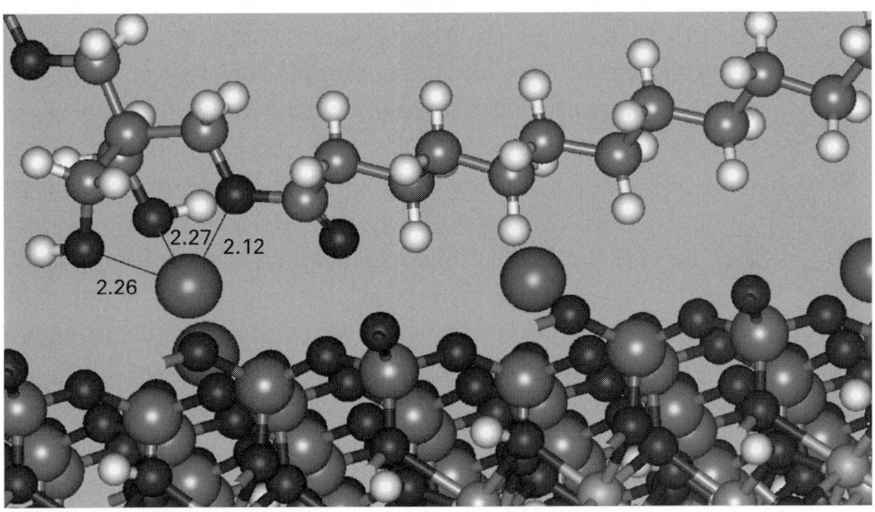

Figure 2.9 Molecular model of a molecule of pentaerythritol distearate bonded to the sodium ion on the clay surface showing two hydroxyls and one ester linkage bonding to the sodium.

mono and diester to the sodium cation on the clay surface is presented in Figures 2.8 and 2.9, respectively. The magnitude of the spacing indicates that the system is not interdigitated. In both the case of DDP and PES, these large *d*-spacings make it easier for the polymer to overcome the clay–clay interactions. An additional advantage of the PES intercalates

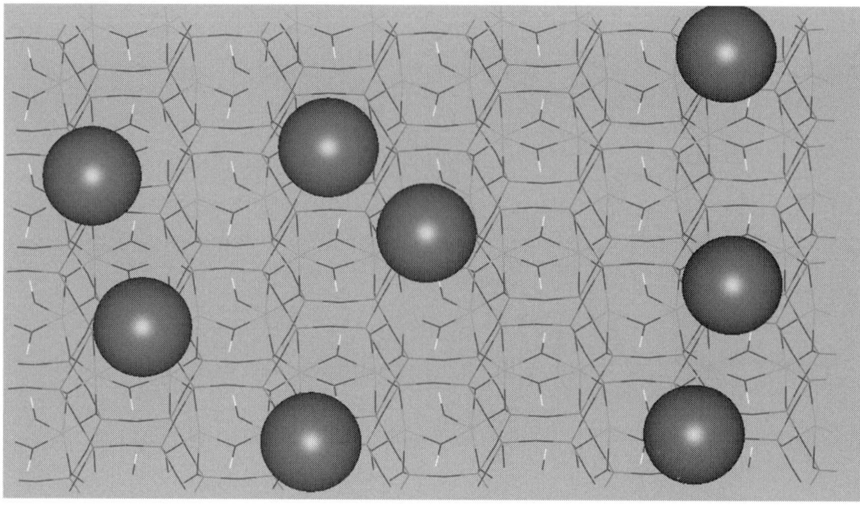

Figure 2.10 Molecular model of 104 meq/100 g montmorillonite illustrating the area of siloxane surface that is available for polymer interaction.

was that the thermal stability was 50–60 °C higher than the dimethyl dehydrogenated tallow treated clay.

Another variation of the ion–dipole surface treatment method utilizing block copolymers has been reported [8]. In this approach, one of the blocks of the copolymer can readily ion–dipole bond to the cations on the clay surface and the other block is compatible with the polymer of interest. Polyethylene oxide is very effective as the block that bonds to the clay surface cations and lies nearly parallel to the clay surface, and the other block is tailored to the polymer of interest and forms loops extending away from the clay surface, better to interact with the polymer.

2.1.9 Hydrophobic polymer intercalation

A second approach to surface modification of the montmorillonite is to take advantage of the siloxane surface that is not occupied by an exchange site. Figure 2.10 represents the surface of a montmorillonite particle that has a cation exchange capacity of 104 meq/100 g. The sizes of the sodium atoms are represented as space filling to illustrate how much area is open on the clay for interaction with the siloxane surface.

This siloxane surface, to all practical purposes, is identical to a talc surface in its behavior. The interaction energy between this surface and a number of polymers has been reported [9,10,11]. In the case of polyacrylamide, the sorption energy was measured to be in the 45 Kj/mol range. In this approach to surface modification, the polymer sorption would be utilized in conjunction with quaternary ammonium or small molecule ion–dipole treatment. This approach has been applied to a thermoplastic polyurethane nanocomposite where the modulus was increased by 260% by a composite made with conventional organoclay but was increased by 300% by a combination of quaternary plus 4% polymer relative to the clay weight [12].

2.1.10 Edge treatment with silane coupling agents

The last method of surface treatment is the use of coupling agents. Unfortunately for montmorillonite, the majority of reactive hydroxides are at the edges of the clay plates. In normal commercially available montmorillonite, these edges account for only about 2% of the total surface area and therefore would not be expected to contribute greatly to the ultimate engineering properties of the composite. This point of view overlooks the fact that the edges are the first part of a clay tactoid that a polymer encounters and therefore can act as a barrier to polymer intercalation, since most polymers do not associate well with hydroxides.

There have been a number of reports where substantial improvement in the nanocomposite physical properties have been demonstrated by using the proper silane coupling agent [13,14,15]. In one of these cases, polypropylene (PP) was shown not to intercalate at all without silane coupling of the edges.

2.2 Thermodynamics of polymer–clay interactions

2.2.1 The enthalpic role in exfoliation

In the previous discussion, a great deal of time has been spent on approaches to surface modification of the clay nanoparticles in order to render the particles to be more compatible with the polymer of interest. This approach mainly concentrates on the enthalpic portion of the Gibb's free energy of intercalation–exfoliation. In order to realize the maximum benefit from a nanocomposite, the exfoliated state is the ultimate goal, since this will present the maximum interfacial interaction between the nanoparticle and the polymer. In reality, a completely exfoliated system probably does not exist, but a Boltzmann distribution of energy states is more likely which invokes some of the entropic terms. In the following discussion, some of the details of the interplay between the entropic and enthalpic energy terms for reaching exfoliation are presented.

The most common view of exfoliation is that if the surface modification of the nanoparticle matches the characteristics of the polymer, then exfoliation should occur. This, however, is too simple a view. A more correct approach is to look at the pairwise interactions between the components of the system.

There have been a number of experimental and theoretical papers dealing with the subject of intercalation–exfoliation [16–19]. These pairwise interactions should include polymer–polymer, polymer–nanoparticle, and particle–particle. The relative strengths of these interactions will lead to different outcomes. The easiest case to deal with is where either the particle–particle or the polymer–polymer interaction energies are so strong that neither intercalation or exfoliation occurs. The second case is where the particle–polymer interaction is very strong. In this case the initial thought would be that this should lead to exfoliation. This case, however, will predominately lead to intercalation rather than exfoliation. The polymer interacts so strongly with the surface of the plates that as it intercalates, it essentially glues the clay plates together, preventing exfoliation.

This type of behavior has been observed for melt intercalation of polystyrene [20,21]. On a theoretical basis, it has been shown that if the polymer has the ability to bond to surfaces from two different plates, only intercalates form and exfoliation is prevented [19]. Where there is very strong interaction between the particle and the polymer, the only case that can potentially lead to exfoliation is one in which the interaction leads to a complex that prevents further bonding to a second plate. This could be a strategy for designing systems to promote exfoliation. In these systems the clay surface could be modified to have a limited number of strongly interacting sites which, once filled, would promote exfoliation. The important balance would be in the number of these sites. If there are too many, then the probability of bonding to sites on opposite plates becomes too large, and if too few, not enough driving force exists to initiate intercalation.

The polymer could also be designed to take advantage of the strong interaction. This could be done in two principal ways. The first would be to design the polymer with a strongly interacting portion at only one end of the molecule. The second would be to design polymers which will bond strongly but, once bonded, could not bond further due to stearic constraints.

An example of this discussed earlier is the comparison of polyethylene oxide (PEO) and PVP. The PEO forms intercalated systems almost exclusively, because it is able to move into position to bond between plates, while PVP, having a more rigid constrained backbone, once bonded to one clay surface places the remaining pyrrolidone groups in positions that make it difficult to bond to a second surface.

Neither will the situation where all pairwise interactions are equal give full exfoliation, since the likelihood of having unintercalated clay, intercalates, and exfoliates will be roughly the same, ignoring entropic contributions. Without a driving force for exfoliation, this situation would yield little exfoliation.

This system could be driven to a higher level of exfoliation by heat and high shear mixing. The ideal situation to yield the highest level of exfoliation is where the particle–polymer interaction is slightly larger than the polymer and clay self-interaction energies. The magnitude of the interaction energy should be on the order of the conformational entropy of a random coiled polymer [22,23].

2.2.2 The entropic role in exfoliation

The entropic part of the Gibb's free energy in forming intercalates and exfoliates is of critical importance. In the classical treatment of entropy, the entropy of mixing should be positive and contribute to the spontaneous formation of intercalates and exfoliated systems. Countering this increase in entropy is the decrease in entropy experienced by the polymer chain, as it must uncoil in order first, to intercalate and, second, to exfoliate the nanoparticle, yielding the maximum interfacial interaction between the polymer and clay nanoparticles.

One exception to this general picture is where the polymer intercalates onto the clay surface and displaces a large number of small molecules. This is exactly what happens in the case of intercalating sodium montmorillonite with PEO or PVP. In these examples, each mole of polymer displaces hundreds of moles of water molecules. The entropy decrease experienced by the polymer in uncoiling is offset strongly by the increase in entropy of the many moles of water displaced from the hydration sphere of the sodium cation on the clay surface.

2.2.3 Kinetics of intercalation–exfoliation

Up until this point, the discussion has centered on the role of thermodynamics in the formation of intercalated and exfoliated polymer clay systems, ignoring any kinetic effects. As can well be imagined, there are a number of activation barriers that must be overcome to allow intercalation to occur. These certainly involve the opening of the clay gallery spacing to allow the polymer to enter, but also, the uncoiling of the polymer to facilitate such insertion.

An additional barrier could also involve the hydroxyl groups studding the edge of the clay platelets. Most polymers will need to overcome the repulsive field presented by these hydroxyls in order to intercalate. As mentioned earlier, this barrier can be reduced or eliminated by treating the edges of the clay with silane coupling agents that now render the edges more hydrophobic in nature.

The intercalation of water-soluble polymers is a clear case where the edge effect does not present any barrier to intercalation. An example would be the intercalation of PVP into sodium montmorillonite [1]. In this case, hydrated sodium montmorillonite as a powder can be mixed with powdered PVP in a bowl mixer and the intercalation occurs in seconds.

A number of studies have been conducted on more hydrophobic polymers to determine the kinetics of intercalation [20,24–30]. The general conclusion of these studies is that the diffusion into the gallery of the clay plates is faster than bulk diffusion of the polymer itself and that the rate of diffusion into the galleries is strongly influenced by the strength of the interaction between the polymer and the surface modification. The relationship is somewhat counterintuitive in that the stronger the interaction, the slower the intercalation. This can be rationalized by the strong interactions almost acting as a sticky surface that slows down the progress of intercalation.

The range of activation energies for intercalation for polybutylene terephthlate, polypropylene, and polystyrene are 15 ± 2, 80 ± 6, and 166 ± 12, respectively. These are not very large activation energies. It would appear that the kinetics of intercalation is not the limiting factor in forming intercalates but that enthalpic interactions dominate. The kinetics of exfoliation at this stage of knowledge is totally unknown. This is mainly due to the difficulty of analytically following exfoliation easily and accurately.

References

[1] G. W. Beall, S. Tsipursky, A. Sorokin, and A. Goldman. *Intercalates and exfoliates formed with oligomers and polymers and composite materials containing same,* US patent number 5552469, 1996-09-03.

[2] R. A. Vaia, S. Vasudevan, W. Krawiec, L. G. Scanlon, and E. P. Giannelis. New polymer electrolyte nanocomposites. Melt intercalation of poly(ethylene oxide) in mica-type silicates. *Advanced Materials*, **7**:2 (1995), 154–156.

[3] D. Siptak. Smectite organoclay chemistry: organically modified bentonite reacted with organic cations. *Chemistry and Manufacture of Cosmetics* (3rd edn) (2002), **3**(Bk 2), 845–855.
[4] P. G. Nahin. Perspectives in applied organoclay chemistry. In *Clays and Oxford, National Clay Minerals, Proceedings of the 10th Conference on Clays and Clay Minerals* (Pergamon, 1963), Austin, Texas, 1961, pp. 257–271.
[5] C. B. Hedley, G. Yuan, and B. K. G. Theng. Thermal analysis of montmorillonites modified with quaternary phosphonium and ammonium surfactants. *Applied Clay Science*, **35**:3–4 (2007), 180–188.
[6] G. W. Beall. The use of organo-clays in water treatment. *Applied Clay Science*, **24**:1–2 (2003), 11–20.
[7] J. Bartels, G. W. Beall, M. Grah, K. Jin, D. Speer, and J. Yarbrough. Intercalated clays from pentaerythritol stearate for use in polymer nanocomposites. *Journal of Applied Polymer Science*, **108**:3 (2008), 1908–1916.
[8] H. Fischer. Polymer nanocomposites: from fundamental research to specific applications. *Materials Science & Engineering, C: Biomimetic and Supramolecular Systems*, **C23**:6–8 (2003), 763–772.
[9] L. T. Chiem, L. Huynh, J. Ralston, and D. A. Beattie. An *in situ* ATR–FTIR study of polyacrylamide adsorption at the talc surface. *Journal of Colloid and Interface Science*, **297**:1(2006), 54–61.
[10] J. C. Wittmann and B. Lotz. Polymer decoration of layer silicates: crystallographic interactions at the polyethylene–talc interface. *Journal of Materials Science*, **21**:2 (1986), 659–668.
[11] V. V. Yatsenko, L. A. Isaenya, M. M. Revyako, and A. Y. Markina. Modification of the surface of fillers in polyethylenimine-based composite materials. *Issled. Obl. Khim. Polietilenimina Ego Primen. Prom-sti.* (1977), 154–5.
[12] E. A. Peterson. Fundamental studies of clay surface treatments to facilitate exfoliation, MA thesis, Texas State University, San Marcos, TX (2005).
[13] M. Kracalik, M. Studenovsky, J. Mikesova, A. Sikora, R. Thomann, C. Friedrich, I. Fortelny, and J. Simonik. Recycled PET nanocomposites improved by silanization of organoclays. *Journal of Applied Polymer Science*, **106**:2 (2007), 926–937.
[14] D.-W. Kim, K.-W. Park, S. R. Chowdhury, G.-H. Kim. Effect of compatibilizer and silane coupling agent on physical properties of ethylene vinyl acetate copolymer/ethylene-1-butene copolymer/clay nanocomposite foams. *Journal of Applied Polymer Science*, **102**:4 (2006), 3259–3265.
[15] S. Kumar and K. Jayaraman. The use of silane coupling agents in polypropylene/clay nanocomposites. In *Proceedings of the 29th Annual Meeting of the Adhesion Society* (2006), pp. 235–236.
[16] K. Kim, L. A. Utracki, and M. R. Kamal. Numerical simulation of polymer nanocomposites using self-consistent mean-field model. *Journal of Chemical Physics* **121**:21 (2004), 10766–10777.
[17] Y. V. Kudryavtsev, E. N. Govorun, A. D. Litmanovich, and H. R. Fischer. Polymer melt intercalation in clay modified by diblock copolymers. *Macromolecular Theory and Simulations*, **13**:5 (2004), 392–399.

[18] B. Chen, J. R. G. Evans, H. C. Greenwell, P. Boulet, P. V. Coveney, A. A. Bowden, and A. Whiting. A critical appraisal of polymer–clay nanocomposites. *Chemical Society Reviews*, **37**:3 (2008), 568–594.

[19] E. Zhulina, C. Singh, and A. C. Balazs. Attraction between surfaces in a polymer melt containing telechelic chains: guidelines for controlling the surface separation in intercalated polymer–clay composites. *Langmuir*, **15**:11 (1999), 3935–3943.

[20] R. A. Vaia, K. D. Jandt, E. J. Kramer, and E. P. Giannelis. Kinetics of polymer melt intercalation. *Macromolecules*, **28**:24 (1995), 8080.

[21] S. Limpanart, S. Khunthon, P. Taepaiboon, P. Supaphol, T. Srikhirin, W. Udomkichdecha, and Y. Boontongkong. Effect of the surfactant coverage on the preparation of polystyrene–clay nanocomposites prepared by melt intercalation. *Materials Letters*, **59**:18 (2005), 2292–2295.

[22] W. Nowicki. Structure and entropy of a long polymer chain in the presence of nanoparticles. *Macromolecules*, **35**:4 (2002), 1424–1436.

[23] W. Nowicki. Properties of systems composed of nanosized particles and very-high-molecular-weight linear polymers. *Seria Chemia (Uniwersytet im. Adama Mickiewicza w Poznaniu)*, **73** (2002), 1–170.

[24] D. Wu, C. Zhou, and H. Zheng. A rheological study on kinetics of poly(butylene terephthalate) melt intercalation. *Journal of Applied Polymer Science*, **99**:4 (2006), 1865–1871.

[25] J. Li, C. Zhou, G. Wang, and D. Zhao. Study on kinetics of polymer melt intercalation by a rheological approach. *Journal of Applied Polymer Science*, **89**:2 (2003), 318–323.

[26] Y. Li and H. Ishida. A study of morphology and intercalation kinetics of polystyrene–organoclay nanocomposites. *Macromolecules*, **38**:15 (2005), 6513–6519.

[27] H. Chen, D. F. Schmidt, M. Pitsikalis, N. Hadjichristidis, Y. Zhang, U. Wiesner, and E. P. Giannelis. Poly(styrene-block-isoprene) nanocomposites: kinetics of intercalation and effects of copolymer on intercalation behaviors. *Journal of Polymer Science, Part B: Polymer Physics*, **41**:24 (2003), 3264–3271.

[28] H. Chen, D. Shah, and E. P. Giannelis. Polymer nanocomposites: interplay between thermodynamics and kinetics. *Polymer Preprints (American Chemical Society, Division of Polymer Chemistry)*, **44**:2 (2003), 243–244.

[29] E. Manias, H. Chen, R. Krishnamoorti, J. Genzer, E. J. Kramer, and E. P. Giannelis. Intercalation kinetics of long polymers in 2 nm confinements. *Macromolecules*, **33**:21 (2000), 7955–7966.

[30] J. Y. Lee, A. R. C. Baljon, D. Y. Sogah, and R. F. Loring, Molecular dynamics study of the intercalation of diblock copolymers into layered silicates. *Journal of Chemical Physics*, **112**:20 (2000), 9112–9119.

3 Analytical methods utilized in nanocomposites

In polymer–clay nanocomposites, to truly reach the ultimate in property improvements requires full exfoliation. A fully exfoliate composite yields the maximum interfacial interaction between the nanoparticle and polymer matrix. In order to produce optimally exfoliated systems requires that direct methods be available to measure the level of exfoliation. The ideal analytical method should be rapid, nondestructive, applicable to many sample matrices, low cost, and should require minimal sample preparation. The only method that fits these criteria is wide-angle X-ray diffraction (WAXD). This method, however, has some major drawbacks that will be discussed in detail in this chapter.

The other analytical methods for confirming the level of exfoliation include scanning electron microscopy (SEM), transmission electron microscopy (TEM), and atomic force microscopy (AFM). The utility and limitations of these three microscopy techniques for measuring exfoliation in nanocomposites will be discussed in detail with specific examples in this chapter.

There are also a number of indirect methods to measure the level of exfoliation but all of them require a direct method with which to standardize them. As examples, two methods, melt viscosity and tensile modulus, will illustrate the indirect methods. Unfortunately, the overall area has not received a great deal of attention, with limited numbers of publications on the subject [1–3].

3.1 Wide-angle X-ray diffraction

In X-ray diffraction the wavelength of the X-rays utilized is of the same order of size as the atomic dimensions in the crystalline lattice and will constructively interfere to yield peaks at discrete angles that can be related to the atomic spacing in the lattice by Bragg's law. The Bragg equation is:

$$n\lambda = 2d \sin \theta$$

where n relates to the order of the reflection, λ is the wavelength of the X-ray employed in the experiment, d is the atomic plane spacing, and θ is the angle of diffraction beam to the atomic plane.

Normally, the experimental variable that is observed is two theta rather than theta due to the geometry of the detector relative to the X-ray source.

As discussed in the previous chapter concerning the structure of smectite clays, it was pointed out that these clays are turbostatic and have WAX patterns that only exhibit 001 or hk0 reflections with no hkl reflections. The hk0 reflections are invariant and can be used to confirm the presence of the clay in a composite at any level of exfoliation. The 001 reflections are a result of the basal spacing between the clay plates as stacks in tactoids. These peaks are highly variable and are dependent upon a number of variables, including level of exfoliation, intercalate type, and level of order. Figure 3.1 illustrates these two types of reflection for montmorillonite intercalated with dodecylpyrrolidone surface treatment where the peak at approximately 20° two theta is of the hk0 type and the peaks at 2.58, 5.26, and 7.84 are of the 001 type. The three peaks are related, being the 001, 002, and 003 reflections.

At first glance, WAX would appear to be the ideal method for determining the level of exfoliation, since it is nondestructive, applicable to a large variety of sample types, relatively inexpensive, and requires comparatively little and easy sample preparation. On initial examination, it should be easy to determine the level of exfoliation simply by watching for the disappearance of the basal spacing peak. This, however, turns out to be much more difficult in practice.

There are several factors that contribute to this difficulty. The first is the fact that the basal d-spacing peaks tend to be rather broad, due to the fact that the intercalated molecules in the gallery tend to pack somewhat randomly, thus yielding a fairly wide distribution of d-values in between the plates. This is further exacerbated by the fact that in polymer composites, the loading of clay nanoparticles is normally 5 wt.% or less. Second, as the gallery begins to open up, diffraction intensity begins to appear at lower angles, tending to obscure the presence of the parent intercalate peak. In addition, as the tactoids of clay begin to exfoliate and go below 100 nm in size, there is significant broadening of the peak. This broadening can be used to calculate a size for the tactoid by using the Scherer equation [4,5]:

$$\tau = K\lambda/\beta \cos\theta$$

where K is a shape factor, λ is the wave length of the X-ray, β is the peak broadening at half maximum intensity (in radians), and θ is

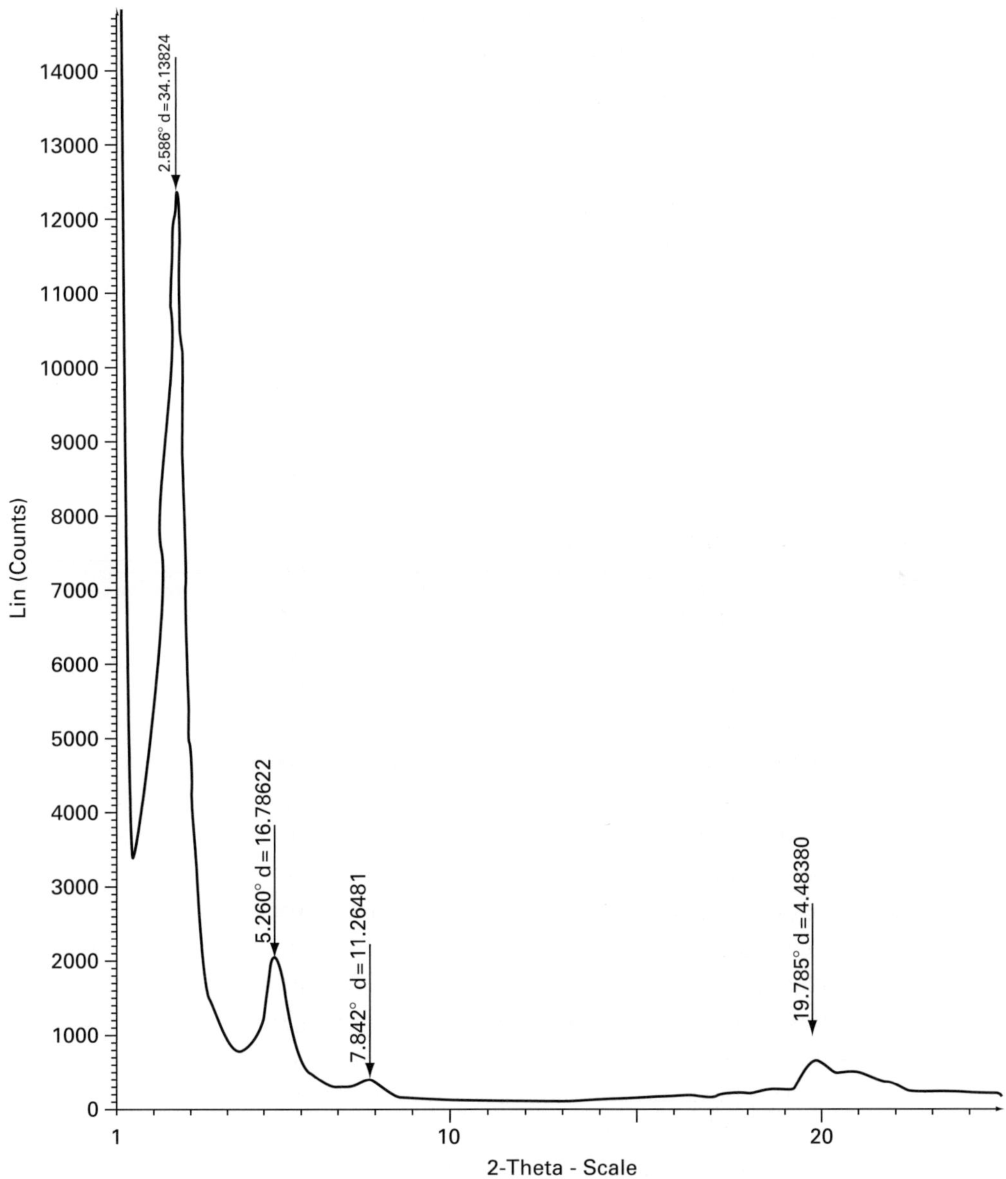

Figure 3.1 X-ray diffraction pattern of montmorillonite intercalated with dodecylpyrrolidone.

the Bragg diffraction angle. This broadening will serve to obscure the basal *d*-spacing even more.

In the worst case, any discrete peaks may disappear, leaving what appears to be a smooth curve, implying an exfoliated system. In this case, however, it is just a mixture of clay tactoids with a continuously varying

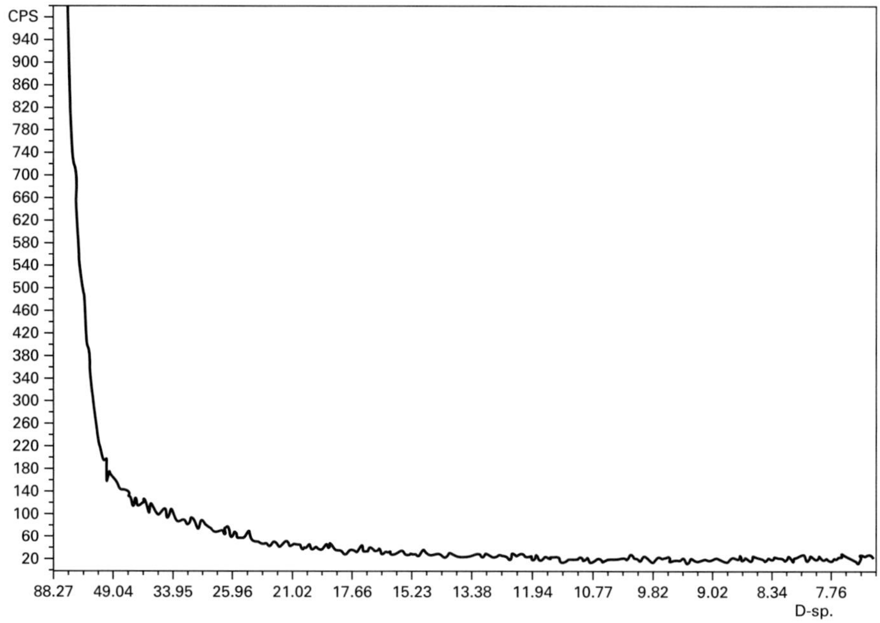

Figure 3.2 X-ray diffraction pattern of a clay–polymer composite that would appear to be exfoliated but actually contains disordered tactoids.

d-spacing that are moving toward exfoliation or very small tactoids. An example of this type of composite can be seen in Figure 3.2, where the diffraction pattern appears to be flat but is simply the superposition of a series of *d*-spaces.

This sort of diffraction pattern can also be exhibited by a system that contains large tactoids that are highly disordered. This can be illustrated by schematic representations as in Figure 3.3, where three montmorillonite plates are disordered relative to one another but still associated in a tactoid. The schematic contains a series of five *d*-spacings that would form the type of diffraction pattern seen in Figure 3.2. It can be seen that in this type of case, the clay plates are still involved in tactoids that can in many cases be very large but are disordered in the sense that there are continuously varying *d*-spaces between plates owing to the misorientation between them. An example of this case will be given later, in the discussion on microscopy methods.

The WAX method therefore appears to be a poor one for quantifying the level of exfoliation. It has, however, the utility of confirming a lack of exfoliation or the presence of intercalated tactoids. This can be enhanced by the method of sample preparation utilized. In the melt, smectite clays have a strong tendency to orient parallel to flow lines. If the X-ray

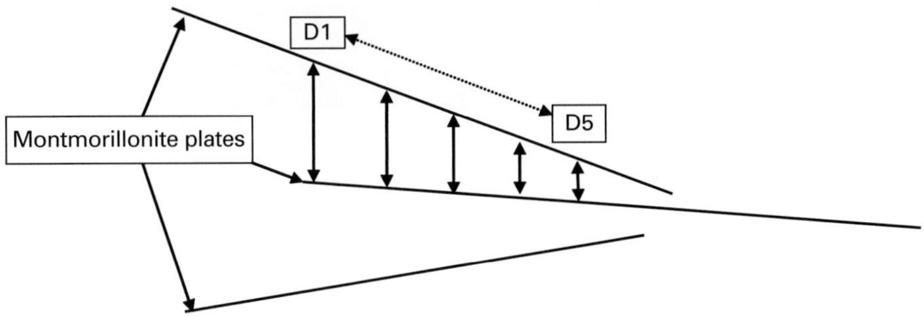

Figure 3.3 Schematic representation of a disordered tactoid illustrating how a series of *d*-spacing would give rise to a featureless X-ray diffraction pattern.

samples are prepared either by extruding through a slit die or by compression-molding the sample, the clay plates will be highly oriented parallel to the film surface. The authors have observed this orientation even in injection-molded bars.

In this case a molded bar of an intercalated nanocomposite was initially X-rayed to confirm the presence of the intercalates. The sample was then sanded to remove a small portion of the surface. This was repeated until the intensity of the basal spacing leveled out. This indicated that only the flow lines due to molding could induce orientation of the tactoids down to a depth of 100 μm. This would strongly enhance the diffraction from the basal planes. If the sample is completely oriented, the hk0 reflection will disappear completely. Using this method, the presence of tactoids can be detected at very low levels. In order to confirm exfoliation, the microscopy methods must be accessed.

3.2 Transmission electron microscopy (TEM)

TEM is a very common analytical technique that can produce high resolution photographs of nanoparticles in polymer matrices. This technique can yield accurate information on the level of exfoliation, the distance between particles, and the spacial distribution of nanoparticles. Figure 3.4 contains the TEM image of a well-exfoliated composite of montmorillonite in nylon 6. The clay plates appear as dark lines in a light gray matrix. The technique can also yield pictures of tactoids and disordered structures mentioned previously when discussing X-ray diffraction.

Figure 3.5 contains an image of a composite that contains large tactoids. The distance between plates in the tactoid can be measured easily and correlated with X-ray diffraction measurements.

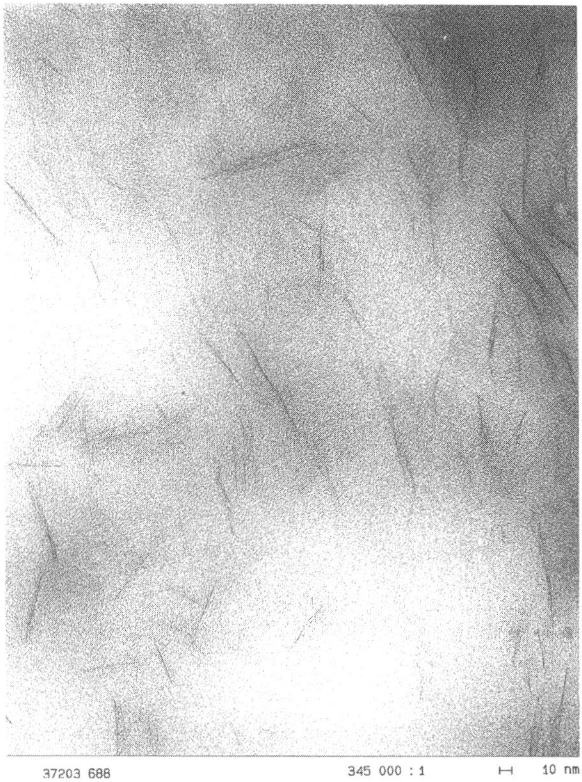

Figure 3.4 TEM image of a nylon 6 clay nanocomposite showing well-exfoliated clay plates.

Figure 3.6 contains an image of a disordered tactoid. It can be seen that the spacing between plates varies widely and the resulting diffraction pattern would be as in Figure 3.2.

Although TEM can yield very good measurements of the levels of exfoliation in a polymer–clay nanocomposite, the technique has a number of difficulties and limitations. The difficulties of the technique revolve mainly around sample preparation. The samples must be very thin, on the order of 100 nm, to allow the electron beam to pass through.

The preparation of very thin samples is normally conducted using a microtome. The samples to be microtomed are normally mounted in an epoxy resin to facilitate the slicing of the sample. For high-modulus polymers, this slicing can be done at room temperature. Many low-modulus and elastomeric samples require cryogenic cooling in order to obtain samples suitable for imaging in the TEM. The very thin and

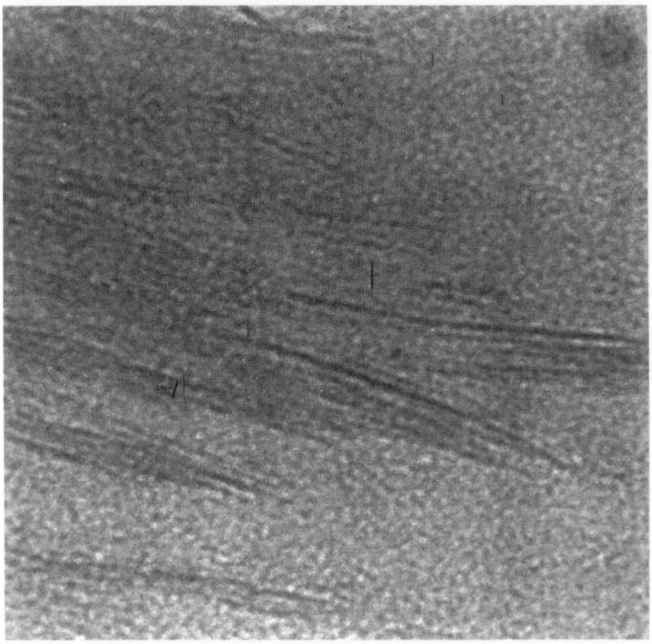

Figure 3.5 TEM of a poor clay–polymer composite with a large tactoid in which the individual *d*-spacing can be measured.

fragile slices are normally floated onto copper grids that hold the sample, allowing for storage and transfer to the microscope for imaging.

The production of samples using this technique is tedious and time-consuming, requiring a high level of experience. The fact that the samples are so thin leads to the limitation that it is difficult to ensure that a representative sample is being imaged. This dictates that multiple random samples of the composite should be imaged.

To increase the quality of images, it is recommended that the nanocomposite be extruded or compression molded into a film in order to orient the clay plates. The film is then mounted in the epoxy such that the microtome slices are cut normal to the surface of the films. This will ensure that the maximum number of plates will be edge on in the TEM image. In a typical X-ray diffraction experiment, the sample can be prepared and run in an hour or two.

TEM imaging can easily require several days. The operation of modern X-ray equipment requires minimal training while sample prep, microtoming, and imaging in a TEM requires extensive training and skill.

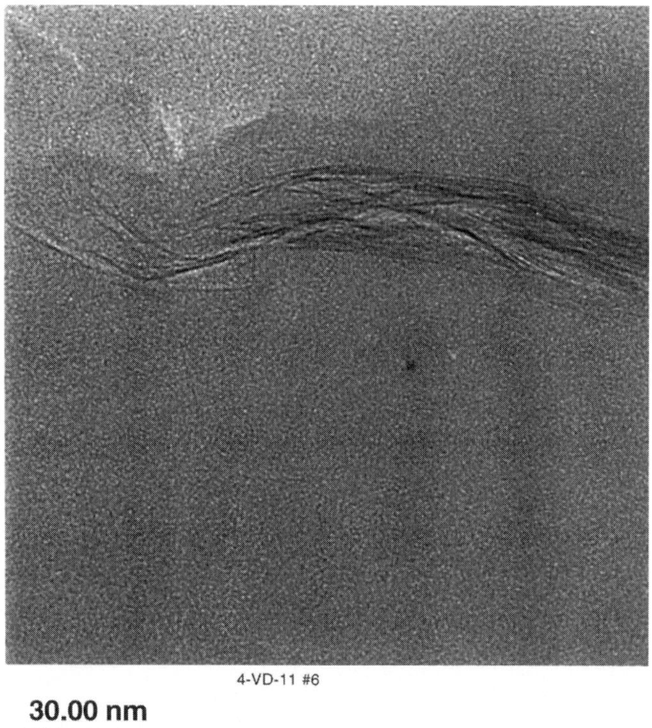

Figure 3.6 TEM of a disordered tactoid that would yield an X-ray pattern that would appear to be exfoliated.

The limitations of the technique relate to two factors: first, to the operator and his/her ability to produce images that are truly representative of the overall sample. In imaging the sample, the tendency is to focus on unusual features in the sample rather than on representative areas. This is a very subjective aspect of the TEM operation. The second limitation relates to measuring distances between plates and the distribution of particles. All of these measurements must be done on a statistical basis and therefore require large amounts of time. Advances in image-processing software have made this task easier but not completely error-free.

TEM is clearly a technique that can produce high-quality images and a measure of exfoliation and tactoid size. The technique, however, is time-consuming, tedious, expensive, and requires a high level of training and skill. It is also limited as to how representative the sampling, imaging, and measurements are for the sample as a whole.

3.3 Scanning electron microscopy (SEM)

SEM is an imaging technique that can yield quality images that are sufficient to yield a good assessment of the level of exfoliation in a nanocomposite. In this technique the electron beam is raster across the surface of the sample being studied. The technique, therefore, is only useful for samples that exhibit topology. This dictates that the nanocomposite must be treated in some way to remove partially the polymer matrix to leave the clay nanoparticles extending from the surface. Attempts to do this etching with solvents are very difficult because surface-tension effects tend to cause the plates to lay down on the surface. Ion beam or plasma etching, however, works very well to etch selectively the polymer, leaving the clay.

Figure 3.7 contains an image of a nanocomposite of montmorillonite in MXD6 polyamide that has been argon-plasma etched. In this image it can be seen that the composite is mostly exfoliated, but some very small tactoids can be seen. These kinds of image can yield a general feel for the level of exfoliation, but again, quantifying the level gives rise to the same problems as described for TEM, but, in addition, the three-dimensional nature of the images adds a level of difficulty. The problems of sampling and imaging representative areas of the sample also exist for SEM. The sample prep difficulties with SEM are not as extensive as with TEM but are not trivial either. The operation and maintenance of SEM is intermediate between X-ray diffraction and TEM.

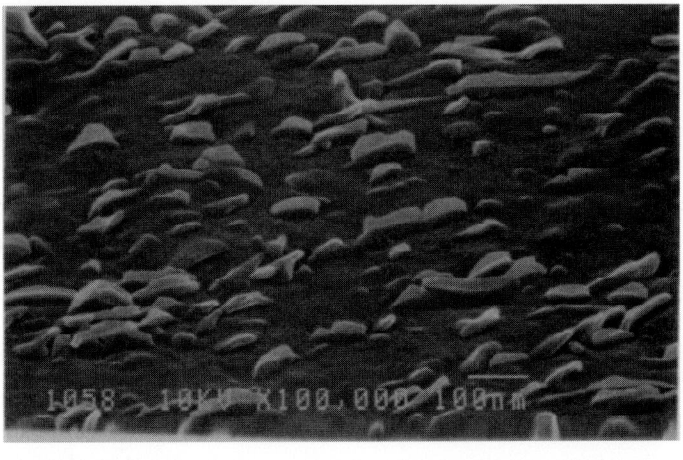

100 nm

Figure 3.7 SEM of a microtomed argon etched surface of MXD6 nanocomposite showing the montmorillonite plates sticking up from the surface.

3.4 Atomic force microscopy (AFM)

AFM is a surface-imaging technique, as is SEM, that is capable of yielding nm-scale resolution. The instrument utilizes a sharp tip mounted on a cantilever that is moved across the sample surface with piezoelectric drivers. The height of the stylus is monitored by reflecting a laser off a mirror mounted on the upper side of the stylus. The stylus can be operated relative to the surface in a great number of modes. In broad terms these modes can be divided into contact and noncontact. In most cases the cantilever is operated to measure the topology of the surface. This means that for nanocomposites, the sample to be imaged must be treated in ways analogous to those described in the section on SEM where the clay nanoparticles must be exposed by removing the polymer.

AFM can, however, be operated in one mode where this kind of sample preparation can be avoided. In this mode the point of the stylus is pressed into the surface with a specific force and the modulus of the material is measured. In this mode the polymer modulus is normally much lower than that measured for the clay particles. Figure 3.8 is an example of the type of image that can be obtained from AFM. In this image the sample has been exposed to a swelling solvent and the constrained polymer around the clay swells less than the unconstrained polymer and, therefore, the clay plates reside at the bottom of the valleys.

3.5 Indirect methods

The methods discussed could be classified as direct methods, since they measure the nanoparticles directly in some way. It is common in analytical measurements to have methods that measure some effect that originates from the true quantity that is of interest. These methods are commonly classified as indirect. In order to yield the highest accuracy in these methods, a direct method must be utilized to correlate the indirect measurement to the quantity of interest.

In polymer nanocomposites, there are two common indirect methods. The first is the measurement of melt viscosity. In this method a rheometer is normally utilized where the viscosity can be measured as a function of shear rate. Clay nanoparticles normally impart thixotropic rheology to the nanocomposite melt. In this type of rheology the melt viscosity at low shear will be much higher than the pure

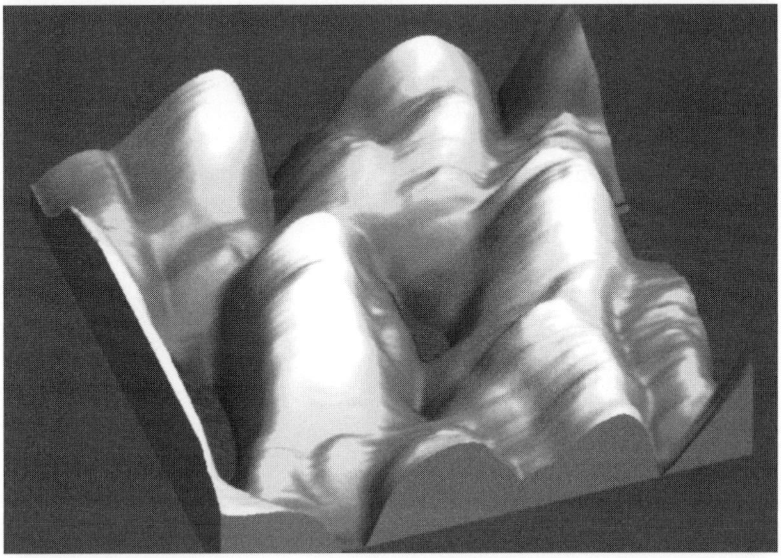

Figure 3.8 AFM image of a microtomed sample of a MXD6 nanocomposite that has been exposed to water to swell it differentially.

polymer, but as more shear is applied, the viscosity will drop to values that are within approximately 10% of that for the pure polymer. The second indirect method would be to monitor the modulus of the composite and utilize the Halpin–Tsai equation to determine the extent of observed reinforcement.

It transpires that the shear rate in most extruders utilized to process polymers is in the high range and so the torque observed in the extruder is not normally markedly different than that of the pure polymer. When the shear rate is lowered, thixotropes undergo hysteresis, with the viscosity returning more slowly than it did when the shear was increased.

References

[1] T. J. Pinnavaia and G. W. Beall (eds). *Polymer/Clay Nanocomposites* (John Wiley & Sons, Inc., Hoboken, NJ, 2001).

[2] W. Michael, A. Elas, K. Berdal, U. Klemradt and D. Carmele. Evaluating dispersion in nanocomposites. *Plastics Research Online, SPE*, July, 15 2009.

[3] A. Reyna-Valencia, Y. Deyrail, and M. Bousmina. *In situ* follow-up of the intercalation process in a clay/polymer nanocomposite model system by

rheo-XRD analyses. *Macromolecules* (Washington, DC, USA), 43:1 (2010), 354–361.

[4] B. D. Cullity and S. R. Stock. *Elements of X-ray Diffraction*, 3rd edn (Prentice-Hall, Upper Saddle River, NJ, 2001), pp. 167–171.

[5] R. Jenkins and R. L. Snyder. *Introduction to X-ray Powder Diffractometry* (John-Wiley & Sons, New York, 1996), pp. 89–91.

4 Gas diffusion characteristics of polymer–clay nanocomposites

4.1 Potential of polymer–clay nanocomposites as barrier materials

Early in the development of polymer–clay nanocomposites it was recognized that, due to the platy morphology of the smectic nanoparticles, the gas permeability of the composite would be altered considerably from that of the pure polymer. This improved barrier has major applications potential in the food and pharmaceutical industries. These composites have the additional advantage of maintaining clarity of display of packaged foods or medicines. The fundamental origin of the barrier properties exhibited by polymer–clay nanocomposites appears to derive largely from the physical morphology of the nanocomposites, but in some notable cases, this cannot be explained by the physical barrier of the nanoparticles.

The number and types of applications utilizing the barrier properties of polymer–clay nanocomposites are significant. In general terms the majority of applications involve the protection of food or drugs from the ingress of either oxygen or water vapor. In the area of flexible food packaging, the nanocomposites will not only protect the food from spoilage and improve shelf life, but also should allow down-gauging in applications where the existing packaging barrier is sufficient. Because of the size and refractive index of the clay nanoparticles, the packaging will also be transparent.

Light-scattering is a function of refractive and diffractive mechanisms. The particles are too small to diffract visible light. The refractive index of clay is very close to the refractive index of polymers utilized in packaging.

For rigid-food packaging, three notable types of food, which are all very sensitive to oxygen, could benefit greatly from these composites. Ketchup currently requires either glass bottles or a five-to-seven-layer plastic bottle.

The second is a related tomato-based food, Salsa. Most Salsa is still sold in glass containers, especially smaller containers, and in complex multilayer containers when sold in larger quantities. The largest of the three is the replacement of the glass beer bottle. In all of these food applications, the use of quaternary ammonium ions to modify the clay will not yield the simplest package, since they are not approved for direct food contact. The modification chemicals described earlier that are classified by the Food and Drug Administration as GRAS will be preferred.

The last very large packaging application is for carbonated drinks. In this case the egress of carbon dioxide rather than ingress of oxygen is the main function of the barrier in the package. In 2L and 3L bottles, an improved barrier could increase the shelf life sufficiently to eliminate the need for discounted sales on shelf stock by extending their effective shelf life. It would also allow much more market penetration for the smaller package sizes.

4.2 Models for gas transport in polymer–clay nanocomposites

4.2.1 The tortuous path model for barrier in nanocomposites

The barrier properties of a number of polymer–clay nanocomposites can be predicted reasonably well by a simple tortuous path model first proposed by Nielsen [1]. In the simple two-dimensional model, the clay nanoparticles act as impermeable barriers to the migration of gas molecules. The equation for the model is:

$$P_c/P_u = V_{fp}/[1 + (L/2W)\ V_{fc}]$$

where P_c is the gas permeability of the nanocomposite, P_u is the gas permeability of the pure polymer, V_{fp} is volume fraction of the polymer in the nanocomposites, *Vfc* is the volume fraction of the clay nanoparticle, *L* is the longest dimension of the clay nanoparticle, and *W* is the smallest dimension of the clay nanoparticle. A schematic representation of the tortuous path model can be seen in Figure 4.1.

In this model the interaction between the clay and polymer is assumed to be minimal and the plates are in a perfectly tiled arrangement parallel to the surface of the polymer film. The *L*/2*W* term for aspect ratio is derived from the assumption that the morphology of the clay particle's cross-section is a rectangle.

Figure 4.2 gives the relative permeability as a function of aspect ratio at a clay loading of 5 wt.%. It can be seen that for most clays which exhibit aspect ratios in the range of 100 to 200, the improvement in

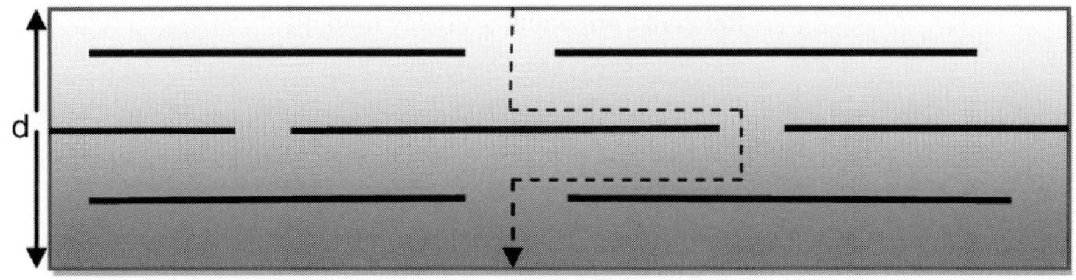

Figure 4.1 Schematic of the tortuous path model showing the increased path due to the high-aspect ratio clay plates.

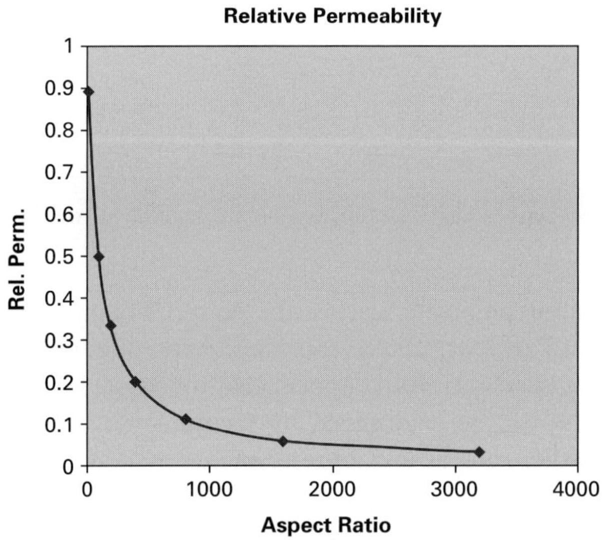

Figure 4.2 Relative permeability of polymer–clay nanocomposites as a function of the clay aspect ratio.

relative permeability is a factor of 2 to 3. Some synthetic fluoromicas and vermiculites have aspect ratios of between 1500 and 3000 and could be expected to lower the gas permeability by factors between 10 and 25. The data in Figure 4.2, as mentioned previously, were calculated at one clay loading.

It is clear from much of the data presented in previous chapters that in the range of 5% clay loadings is the point of diminishing return in mechanical performance improvements for thermoplastics. Although the aspect ratio strongly affects the relative permeability, it is also important to look at the effect of clay loading.

Figure 4.3 is a plot of relative permeability for nanocomposites at five different aspect ratios as a function of clay loading. It is clear that the

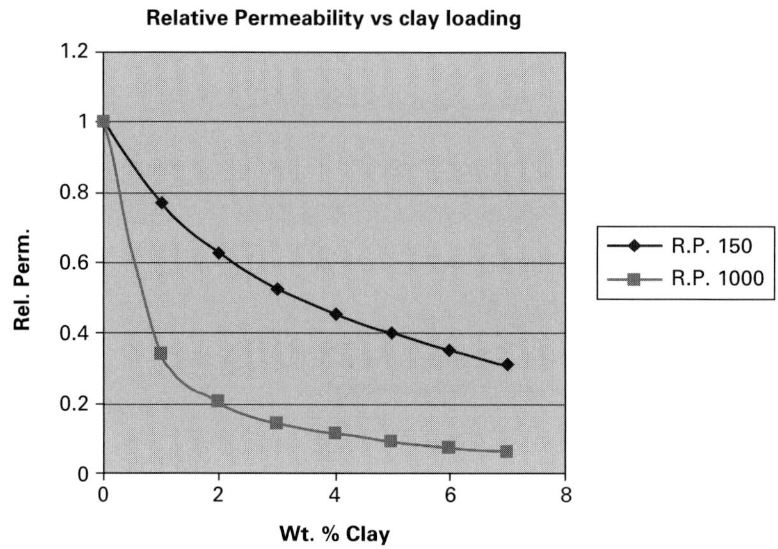

Figure 4.3 Relative permeability as a function of clay loading at two different aspect ratios.

1000 aspect ratio composite reaches the point of diminishing return at low loadings of 2 to 3 wt.%. The 150 aspect ratio does not start leveling off until above 7 wt.%. It would appear that for most clays, the practical limit of loading will be determined by factors such as film clarity, or physical properties such as impact resistance.

4.2.2 Experimental data on nanocomposite barrier performance

The experimental data on barrier performance of polymer–clay nanocomposites can be divided into three categories. The first group includes systems that fit the tortuous path model reasonably well. The second category includes composites that exhibit relative gas permeabilities that are inferior to the tortuous path model predictions. The last group includes composites that exceed the performance predicted by the model. The following discussion will give specific examples of each of these categories with explanations for each type of behavior.

4.2.2.1 Examples of nanocomposites that fit the tortuous path model

Examples of nanocomposites that fit the tortuous path model relatively well include work by Shah *et al.* [2] with low-density polyethylene (LDPE) and ionomer, and by Gatos and Karger-Kocsis [3] with a nanocomposite of nitrile butyl rubber (NBR) and organofluorohectorite

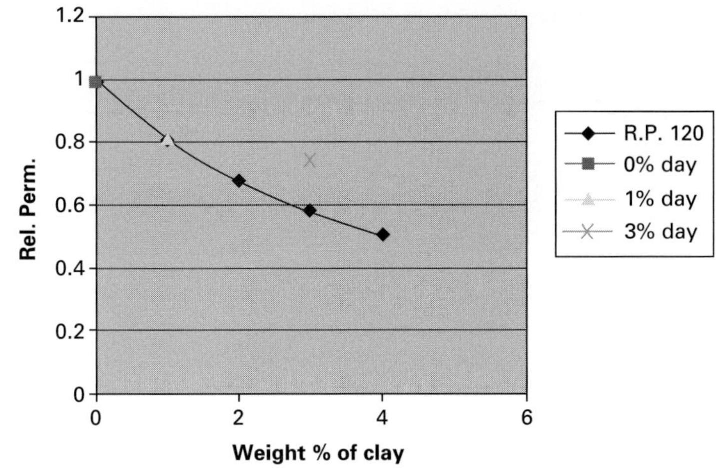

Figure 4.4 Relative permeability for LDPE nanocomposites compared to the tortuous path model for an aspect ratio of 120.

nanoparticles. The study by Shah evaluated the relative permeabilities for nitrogen, oxygen, and carbon dioxide. The permeability of the gases is a function of the solubility of the gases in the polymer nanocomposites and the rate of diffusion. The rate of diffusion will be a function of the size of the gas molecules. Because nitrogen is larger than oxygen and the solubilities are approximately the same [14], the permeability of oxygen was faster than nitrogen through the polymer nanocomposites.

The relative gas permeability of nitrogen for nanocomposites of LDPE compared to the predicted curve using Nielsen's model with an aspect ratio of 120 is plotted in Figure 4.4. At 1 % clay loading, the agreement is very good, but at 3 %, there is a significant negative deviation from the predicted curve. This deviation at higher loadings is normally an indication that full exfoliation in the composite has not been realized.

The second polymer studied by Shah was an ionomer (Surlyn 8945). Figure 4.5 contains the relative permeabilities for the ionomer nanocomposite compared to the tortuous path model predictions for an aspect ratio 120. The agreement between the tortuous path model and experimental is quite good. The generally accepted aspect ratio for this particular clay is about 150, which is close to the 120 utilized in the figures. The clay in the ionomer appears to be more fully exfoliated, which is consistent with the higher percent increase in modulus measured for the ionomer nanocomposite relative to the LDPE.

The next example is a nanocomposite of NBR with a fluorohectorite. In this composite the transmission electron microscopy (TEM) and

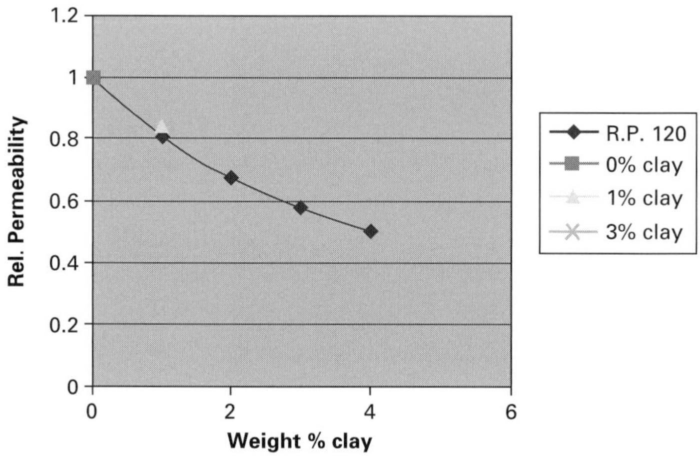

Figure 4.5 Relative permeability for ionomer nanocomposite compared to the tortuous path model for an aspect ratio of 120.

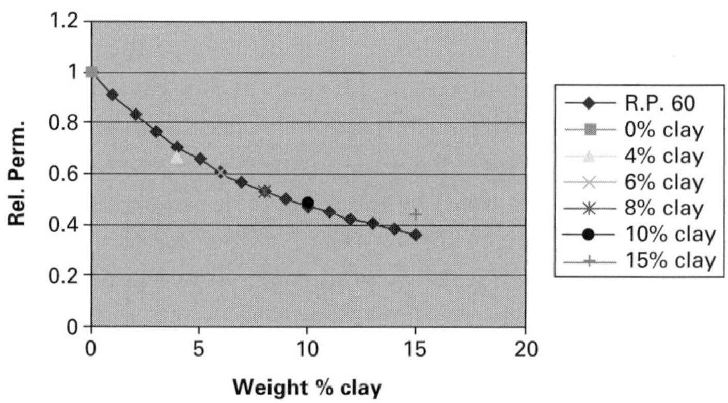

Figure 4.6 Relative permeabilities for NBR nanocomposites at various clay loadings compared to the tortuous path model with an aspect ratio of 50.

wide-angle X-ray diffraction shows that the system was tactoidal in structure, with the tactoids having an aspect ratio of about 50. The individual plates of the fluorohectorite have an aspect ratio of approximately 200. This would indicate that the tactoids have approximately four plates per tactoid. Figure 4.6 contains a plot of relative permeability for this composite compared to the predicted curve for the tortuous path model for an aspect ratio of 50. The agreement is very good between the experimental and calculated data. It is important to point out that this agreement is over a large concentration range of 0–15 wt.%. This is a very good example of how to account for tactoidal structure in the aspect

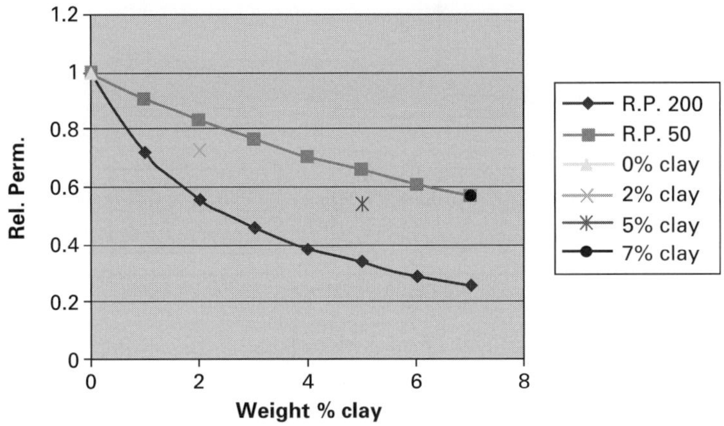

Figure 4.7 Relative permeability for an EVA nanocomposite compared to the tortuous path model using aspect ratios of 200 and 100.

ratio calculation utilizing data from TEM. The type of agreement seen in these three examples applies to a considerable number of polymer–clay nanocomposites.

4.2.2.2 Nanocomposites that exhibit negative deviations from the tortuous path model

A second group of nanocomposites mentioned earlier is characterized by exhibiting negative deviations from the tortuous path model. These account for another significant number of composites. In these composites this negative deviation can be explained by one of two possible mechanisms. The first is the one mentioned in the previous example where tactoids predominate in the composite. In these cases the aspect ratio of the tactoids has not been determined but the data is compared to the tortuous path equation predictions utilizing the aspect ratio of the completely exfoliated clay. The solution in these cases is to utilize X-ray diffraction and electron microscopy to determine the true aspect ratio of the tactoids. This was done in the previous example.

The experimental data utilizing the measured aspect ratio should be the basis for comparison. An example of this type of composite is data reported by Zhong *et al.* [4] on a nanocomposite of ethylene vinyl acetate (EVA) with montmorillonite. Figure 4.7 contains the experimental relative permeabilities for this composite compared to the curves predicted by the tortuous path model utilizing aspect ratios of 200 and 50. The 200 aspect ratio represents the fully exfoliated nanoclay and the 50 represents

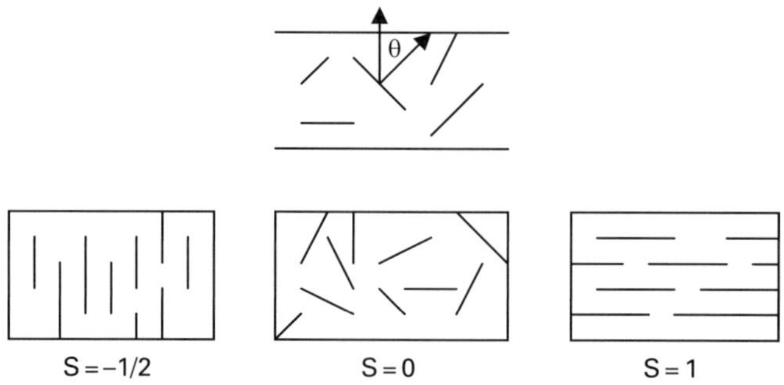

Figure 4.8 Schematic diagram of the angle θ between the normal to the film surface and the normal to the clay plate and several examples of the orientation factor S.

tactoids that contain two clay layers. The best fit is closer to a tactoidal structure containing approximately two clay layers per tactoid.

The second explanation for the deviation involves the randomization of the arrangement of the clay particles in the polymer nanocomposite, a disorder factor. This disorder factor attempts to account for misalignment of the clay nanoparticles that lowers the cross-section that a diffusing molecule encounters traversing the composite film. This approach has been applied to the Nielsen equation by Bharadwaj [5].

Bharadwaj invoked an orientation factor that accounts for deviations of the plates from perfect parallel alignment with the surface of the nanocomposite film.

The orientation factor S is defined as:

$$S = 1/2(3\cos^2\theta - 1),$$

where θ is the angle between the vector **n**, that is normal to the film surface and the vector, **p**, which is normal to the clay plate surface. Figure 4.8 contains a schematic that illustrates these vectors and several values for S. The orientation factor is applied to the tortuous path model in the following way:

$$P_c/P_u = V_{fp}/(1 + \{[(\mathrm{L}/2\mathrm{W})\ \mathrm{V}_{\mathrm{fc}}](2/3)(\mathrm{S} + 1/2)\}$$

The general effect of the orientation factor is to lower the effective aspect ratio. When $S = 1$, the equation reduces to the Nielsen equation. Fredrickson and Bicerano [15] applied the same logic with nanoparticles that were modeled as disks. An excellent review of these models, including the Cussler–Aris model [16], is provided by Manias *et al.* [17].

The Nielsen equation above with the orientation factor correction is preferred for polymer–clay nanocomposites. The assumption by the

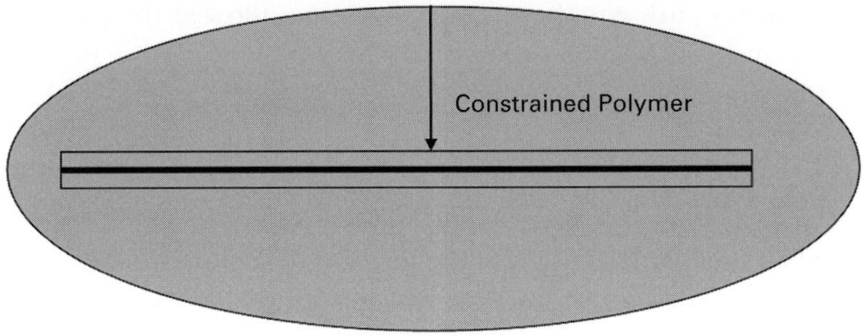

Figure 4.9 Schematic diagram of the constrained polymer region around the clay plate.

Nielsen model of a square morphology for clay is not precisely applicable. Work done by Rouse *et al.* [18] indicates that clay similar to that utilized to prepare Cloisite 6A for the work done by Shah *et al.* [2] above has a plate structure, but is not square. They appear to be like tattered sheets. This complex morphology is not easily accommodated into simple equations describing the Nielsen model. The size range is variable. The micrographs indicate particles in the range of 200 to 600 nm. Many smaller particles are also present.

4.2.2.3 Nanocomposites that deviate positively from the tortuous path model and the constrained polymer model

The last class of nanocomposites includes a group that exhibit relative gas permeabilities that exceed the performance predicted by the tortuous path model. This behavior would indicate that the nanoparticles have altered the fundamental behavior of the polymer.

It has been reported by Lin *et al.* [6] that the diffusion coefficient of very thin polymethylmethacryalate polymer films interacting strongly with a substrate can be two orders of magnitude lower than thick films. This effect extends from the surface of the substrate approximately 40 nm.

Beall [7] proposed that this effect was important in forming a constrained polymer region around the platy clay nanoparticles. This region is responsible for deviations from the tortuous path model in this class of nanocomposites. In this model, two parameters were unknown. The first was the size of the constrained region and the second was the diffusion coefficient within the constrained region. This model is illustrated in Figure 4.9. Since this was proposed as a conceptual model, the effect of change in the diffusion coefficient and the size of the constrained region were lumped into one correction factor applied to the numerator of the tortuous path equation. This correction allowed the prediction of both the magnitude and shape of the curves for relative permeability as a function of clay loading.

4.2.3 Data supporting the constrained polymer model

The fundamental origin of the lowered diffusion rates in this constrained region is related to lowered solvent uptake due to the polymers being restricted in their ability to move. Restriction of polymer motion is a function of the interaction of the polymer with the surface of the nanoparticle. This proposed mechanism is supported by observations of solvent uptake [8,9,10,11,12] made on a number of nanocomposites. In each of these cases the solvent uptake decreased with increasing nanoparticle loading and was substantially lower than the solvent uptake of the pure polymer.

The large changes observed in solvent uptake in these nanocomposites provide strong indirect evidence of the constrained polymer region and its size. In the work on solvent uptake by ethylene-vinyl acetate nanocomposites, the solvent uptake effect levels off at around 5 wt.% clay nanoparticles. This indicates that nearly all the polymer in the composite is constrained at this loading level. At 5 wt.%, assuming that the clay plates are evenly dispersed, the distance between plates would be approximately 50 nm. The constrained polymer region would be required to extend at least 25 nm from the surface.

The size of the constrained polymer region has been directly measured by Beall [13]. In this study, two polyamide/clay nanocomposites were studied, nylon 6 and MXD6. The nanocomposites were melt-compounded and extruded as thin sheets. This will strongly orient the clay nanoparticles parallel to the surface of the film. These films were then microtomed normal to the film surface and mounted on silicon wafers. The two polyamide/clay nanocomposite samples were initially scanned after drying in an AFM to confirm that no topographic features existed. The samples were then exposed to a high humidity atmosphere and allowed to swell. The assumption was that the constrained region would swell less than the unconstrained. After swelling the MXD6–clay samples were again scanned in the AFM. Large hills and valleys appeared. The composite containing 0.5 wt.% clay had valleys that were approximately 200 nm in length. This is very close to the length of the clay plates in the composite. The plates were measured to be over 500 nm apart, indicating that the constrained regions, on average, are not

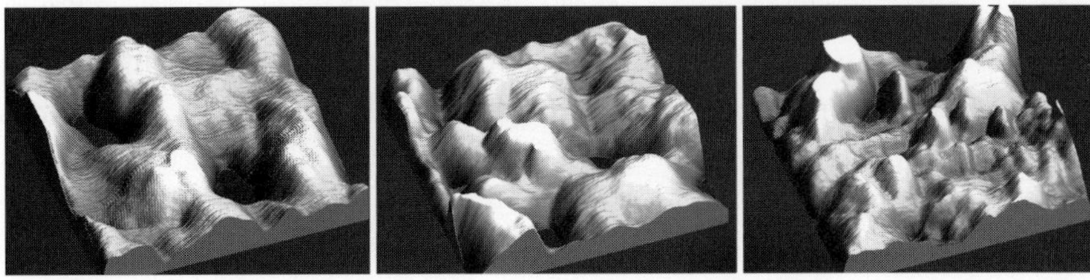

Figure 4.10 AFM pictures of the topography of a swelled MXD6 nanocomposite at various clay loadings showing valleys and hills owing to constrained polymer regions.

overlapping. The average distance across the valley, near what appears to be the center of the clay plate, was 82 nm, which would mean that the constrained polymer region extends approximately 40 nm from the clay surface. As the clay loading increases from 0.5 to 2.0 to 5.0, the number and size of hills decrease. At 5% loading, the distance across the valleys is reduced to 63 nm, indicating substantial crowding of the constrained regions. This would yield a distance of approximately 30 nm for the distance by which the constrained region extends from the clay surface. This is in excellent agreement with the calculated value of 25 nm from solvent uptake.

Figure 4.10 contains AFM images of swelled samples of MXD6 nanocomposites at clay loadings of 0.5, 2.0, and 5.0 wt.% clay. At loadings of 5% or above, the diffusion coefficient for the constrained region could be calculated by using the tortuous path model. This would deconvolute the effect of tortuous path and diffusion in the constrained region. Armed with the size of the constrained region and the relative diffusion coefficients of the constrained and unconstrained regions, the full relative permeability curve can be predicted. When the nylon 6 nanocomposite samples were swelled, the AFM indicated no valleys and hills. Nylon 6 nanocomposites are highly crystalline. The only difference between the pure polymer and the nanocomposite is the crystalline form.

It would be highly unlikely that these different crystalline forms would have radically different solvent uptake behavior. In the case of MXD6, which is amorphous, the effect of constrained polymer could radically alter the solvent uptake. In general, the radical changes in solvent uptake have mainly been observed in amorphous polymers. With further study, it may turn out that only amorphous polymers will exhibit this constrained polymer effect on solvent uptake and large increases in barrier.

NBR, discussed above, is an elastomer utilized in applications where barrier properties are a critical performance criteria. Another elastomer where barrier performance is critical is butyl rubber. Butyl rubber is employed as the inner liner for tire manufacture. Butyl rubber has an excellent gas barrier performance. However, the polymer is the most expensive polymer used in tires, mechanical performance is relatively poor, and the density is high compared to the other polymers utilized in tire manufacture. Improving the gas barrier properties of butyl rubber with clay allows for a decrease in the amount of butyl rubber needed and improves the mechanical properties. Enhancement of mechanical properties is discussed in Chapter 5.

All the work on barrier performance of butyl rubber–clay nanocomposites has been done by industry and is available in the literature as patents and patent applications [19–28]. Unfortunately, the formulas that were prepared for evaluation were patterned on commercial products. Improved barrier performance is recorded but systematic evaluations were not performed. Therefore, an examination of the applicability of the models provided above cannot be carried out. A systematic evaluation of the barrier mechanism of clay in butyl rubber has yet to be done.

References

[1] L. E. Nielsen. Models for the permeability of filled polymer systems. *J. Macromol. Sci.*, **A1**:5 (1967), 929–942.

[2] R. K. Shah, R. K. Krishnaswamy, S. Takahashi, and D. R. Paul. Blown films of nanocomposites prepared from low density polyethylene and a sodium ionomer of poly(ethylene-co-methacrylic acid). *Polymer*, **47** (2006), 6187–6201.

[3] K. Gatos and J. Karger-Kocsis. Effect of the aspect ratio of silicate platelets on the mechanical and barrier properties of hydrogenated acrylonitrile butadiene rubber (HNBR)/layered silicate nanocomposites. *European Polymer Journal*, **43** (2007), 1097–1104.

[4] K. Ghosal and B. D. Freeman. Gas separation using polymer membranes: an overview. *Polymers for Advanced Technologies*, **5**:11 (1994), 673–697.

[5] Y. Zhong, D. Janes, Y. Zheng, M. Hetzer, and D. Kee. Mechanical and oxygen barrier properties of organoclay–polyethylene nanocomposite films, Poly. *Engineering Science*, **19** (2007), 1101–1107.

[6] R. K. Bharadwaj. Modeling the barrier properties of polymer-layered silicate nanocomposites. *Macromolecules*, **34** (2001), 919–9192.

[7] G. H. Fredrickson and J. Bicerano. Barrier properties of oriented disk composites. *Journal Chemical Physics* **110** (1999), 2181–2188.

[8] E. L. Cussler, S. E. Hughes, W. J. Ward, and R. Aris. Barrier membranes. *Journal of Membrane Science*, **38** (1988), 161–174.

[9] E. Manias, G. Polizos, H. Nakajima, and M. J. Heidecker. Fundamentals of polymer nanocomposite technology. In *Flame Retardant polymer nanocomposites*, ed. A. B. Morgan and C. A. Wilkie (John Wiley & Sons, New York, New York, USA 2007), pp. 56–60.

[10] J. H. Rouse, S. T. White, and G. S. Ferguson. A method for imaging single clay platelets by scanning electron microscopy. *Scanning*, **26**:3 (2004), 131–134.

[11] E. K. Lin, Wen-Li Wu, and S. K. Satija. Polymer interdiffusion near an attractive solid substrate. *Macromolecules*, **30** (1997), 7224–7231.

[12] G. W. Beall. New conceptual model for interpreting nanocomposite behavior. In *Polymer–Clay Nanocomposites*, eds T. J. Pinnavaia and G. W. Beall (Wiley, chichester, UK, 2000).

[13] A. Singh and M. Mukherjee. Effect of polymer–particle interaction in swelling dynamics of ultrathin nanocomposite films. *Macromolecules*, **38** (2005), 8795–8802.

[14] M. Promanik, H. Acharya, and S. K. Srivastava. Exertion of inhibiting effect by aluminosilicate layers on swelling of solution blended EVA/clay nanocomposite. *Macromolealor Materials and Engineering*, **289** (2004), 562–567.

[15] S. D. Burnside and E. P. Giannelis. Nanostructure and properties of polysiloxane-layered silicate nanocomposites. *Journal Polymer of Science: Part B: Polymer Physics*, **38** (2000), 1595–1604.

[16] Z.-M. Liang, J. Yin, J.-H. Wu, Z.-X. Qiu, and F.-F. He. Polyimide/montmorillonite nanocomposites with photolithographic properties. *European Polymer Journal*, **40** (2004), 307–314.

[17] Z.-M. Liang and J. Yin. Poly(etherimide)/montmorillonite nanocomposites prepared by melt intercalation. *Journal of Applied Polymer Science*, **90** (2003), 1857–1863.

[18] G. W. Beall and D. Adame. Direct measurement of the constrained polymer region in polyamide/clay nanocomposite. *Applied Clay Science* (submitted 2007).

[19] A. H. Tsou and A. J. Dias. Low permeability nanocomposites, PCT WO 02/100923 A2, 2002.

[20] N. Yagi, K. Muraoka, Y. Minagawa, and K. Nishioka. Pneumatic tire, US patent. 2004/0226643 A1, 2004.

[21] K. Ishida and K. Masaki. High gas-barrier rubber compositions for inner liners and their pheumatic tires, Japan. *Kokai Tokkyo Koho* 2003–10621 20030120, 2004.

[22] C. Gong, A. J. Dias, A. H. Tsou, B. J. Poole, and K. R. Karp. Functionalized elastomer nanocomposites with improved air barrier properties for tire innerliners and innertubes, PCT WO 2004005388 A1, 2004.

[23] C. Gong, A. J. Dias, A. H. Tsou, B. J. Poole, and K. R. Karp. Functionalized elastomer nanocomposites with improved air barrier properties for tire innerliners and innertubes, PCT WO 2004005387 A1, 2004.

[24] K. Hagiwara, K. Kadota, and S. Mashimo. Japan Kokai Tokkyo Koho 2002–130740 20020502, 2002.

[25] T. Maruyama, K. Ishikawa, N. Amino, and M. Ikawa. Organically modified layered clay as well as organic polymer composition and tire inner liner containing same, US patent. 6908958 B2, 2005.

[26] K. Muraoka, N. Yagi, K. Nishioka, and N. Osaki. Tubeless tire, US patent. 2005/0098252 A1.

[27] K. Igawa and K. Ishikawa. Butyl rubber compositions with improved workability and processability in unvulcanized states without impairing air impermeability for inner liners of pneumatic tires. Japan Kokai Tokkyo Koho 2004–59013 20040303, 2005, okai Tokkyo Koho 2004–59013 20040303, 2005.

[28] K. Ishida and K. Fujiki. Rubber composition for inner liner and tire, US patent. 2006/0142463 A1, 2006.

5 Engineering properties of polymer–clay nanocomposites theory and theory validation

5.1 Mechanics

An understanding of the structure–property relationships of polymer–clay nanocomposites during mechanical testing requires an understanding of morphology of the polymer continuous phase at the molecular level, the effect of the clay nanoparticle on the polymer morphology, the interaction of the polymer molecules at the interface of the clay, and the physics of the test for mechanical properties in relation to the changes of the polymer–clay nanocomposite during the test.

For example, if a polymer–clay nanocomposite is being pulled apart by an Instron-type stress–strain tester, there should be a linear relationship between the stress applied and strain (stretching) of the sample below the elastic limit (Hooke's law). The ratio of tensile stress over tensile strain is referred to as Young's modulus. Tensile stress is defined as the force applied to the sample (F) divided by the cross-sectional area of the sample (Area). The strain is defined as the change in the length of the sample (δL) divided by the original length of the sample before stretching (L). Hence, Young's modulus (Y) can be defined as:

$$Y = \frac{F/\text{Area}}{\delta L/L}. \tag{5.1}$$

As the sample is stretched, it changes shape. As the length of the sample increases with stretching, the thickness (T) (and hence the cross-sectional area) will also change (δT). The ratio of the changes in dimension is referred to as Poisson's ratio (P):

$$P = \frac{\delta T/T}{\delta L/L}. \tag{5.2}$$

These relationships work well for isotropic materials. However, many of the commercial polymers that are utilized for engineered applications are not anisotropic at the molecular level [1,2]. Clay is a very anisotropic plate with a thickness of about 1 nm and minimal plate dimensions of 150 to 200 nm [3]. The plates are not regular and can vary significantly in size [4]. The surface area of montmorillonite (a clay utilized in many polymer–clay nanocomposite studies) is in excess of 750 m^2/g [4]. At this surface area, a large polymer molecule can associate exclusively on one montmorillonite particle.

The nature of the polymer–clay interfacial association can be a very significant variable that relates to the mechanical properties of these composites. The complexity of the polymer–clay nanocomposite presents significant challenges in deconvoluting the variables that relate to structure–property relationships as a function of Young's modulus and the determination of Poisson's ratio. These challenges are being addressed today as they relate to Young's modulus and other important mechanical tests.

5.2 Proper preparation and analysis of polymer–clay nanocomposites

An excellent example of the proper protocol to employ in regard to the assessment of the mechanical properties of polymer–clay nanocomposites is provided by the work of Fornes and Paul [5]. The Young's modulus of polymer composites formed by symmetrical, large, reinforcing dispersed phases in polymers where the adhesion of the polymer to the dispersed phases is large can be approximated by:

$$Y_c = Y_p \Phi_p + Y_f \Phi_f, \tag{5.3}$$

where Y_c is the Young's modulus of the composite, Y_p is the Young's modulus of the polymer, Y_f is the Young's modulus of the reinforcing dispersed phase, Φ_p is the volume fraction of polymer in the composite, and Φ_f is the volume fraction of reinforcing dispersed phase. Equation 5.3 can be viewed as describing a mechanical model of pulling two springs. One spring represents the polymer and the other represents the reinforcing dispersed phase. The springs are pulled in parallel.

Clay is very anisotropic and small with a large surface area and modulus. To account for these characteristics of clay in a polymer, Fornes and Paul evaluated two analytical methods, those of Halpin–Tsai [6,7,8] and Mori–Tanaka [9], that were developed to calculate the Young's moduli for the types of morphology that can be associated with clay particles. Assuming that the polymer and the reinforcing dispersed phase are the only components in the polymer composite, the volume fraction of polymer can be expressed as a function of the volume fraction of the dispersed phase:

$$\Phi_p = 1 - \Phi_f. \tag{5.4}$$

Substituting Equation 5.4 into Equation 5.3, Young's modulus can be expressed as:

$$Y_c = Y_p(1 - \Phi_f) + Y_f\Phi_f. \tag{5.5}$$

The expression for the Halpin–Tsai equation that is applicable to polymer–clay nanocomposites that have the clay particles oriented in the direction of the applied stress includes the aspect ratio (A) of the particle in the equation.

5.3 Theory of anisotropic dispersed-phase reinforcement of polymers

Halpin and Kardos [7] published a review that described the derivation of the Halpin–Tsai equation from the first-principle arguments of Hill [10,11,12], Hermans [13], and Kerner [14] that were focused on polymers that are reinforced by dispersed-phase fibers. The Halpin–Tsai Equation (5.6):

$$\frac{Y_c}{Y_p} = \frac{1 + 2A\left[(Y_f/Y_p - 1)/(Y_f/Y_p + 2A)\right]\Phi_f}{1 - \left[(Y_f/Y_p - 1)/(Y_f/Y_p + 2A)\right]\Phi_f} \tag{5.6}$$

was shown to be very robust through derivations from all of the first-principle arguments listed above. The aspect ratio (A) in the Halpin–Tsai equation for clay particles is the diameter/thickness of the clay particle. An implication of the Halpin–Tsai equation in relation to the first-principle arguments derived for fibers is that changing the morphology of the dispersed phase from a fiber to a plate (clay particle) results in reinforcement orthogonal to the direction of applied stress, resulting in an additional increase in modulus. The Halpin–Tsai equation is also sensitive to any anisotropic nature in the polymer (crystalline structures) through Hermans' arguments.

5.4 Genesis: anisotropic dispersed-phase reinforcement of metal alloys

The Mori–Tanaka model is derived from a model developed by Eshelby [15]. The terms in Eshelby's arguments are predicated in the field of metallurgy, hence, they may seem foreign to investigators in polymer composites. The materials discussed in Eshelby's arguments are assumed to obey Hooke's law (the force necessary to change the shape of a material is directly proportional to the change in shape. Pulling on a metal spring is an excellent example of this). This assumption is appropriate for studying metal materials.

Mechanical examination of the composite in Eshelby's discussion involves removal of the inclusion (the dispersed phase) from the matrix (the continuous phase) and operation (stress resulting in strain) on this dispersed phase without the dispersed phase being present in the continuous phase. The dispersed phase is brought back to the same geometry it had when it was present in the continuous phase. The dispersed phase is placed back into the continuous phase, being cognizant of the surface (S) describing the interface between the dispersed phase and the continuous phase. The final step in the process is to let the composite relax. Now, no external stress is being applied to the composite, but internal stress is present because of the change in the dispersed phase.

Through this exercise and careful mechanical analysis, Eshelby derived expressions for the composite to determine the following:

- the elastic field divorced from the dispersed phase;
- the stress and strain immediately outside the dispersed phase in the continuous phase;
- total strain in the composite;
- the relationship between the elastic nature of the dispersed phase and an external elastic field;
- an elastic field divorced from the dispersed phase;
- stress concentrations associated with the surface of the dispersed phase;
- the energy of interaction between the dispersed phase and an elastic field;
- the elastic constants associated with a composite that has a low concentration of dispersed phases of different shapes.

One special case that Eshelby analyzed was that for the morphology of the dispersed phase being a disk. This special case will be important later in this chapter. Mori and Tanake's arguments are also predicated from a metallurgical perspective. Utilizing Eshelby's rationale, they demonstrate that the average stress at any point in the continuous phase of a composite is uniform. The fluctuating stress from the dispersed phase can be averaged to zero.

5.5 Transition from anisotropic dispersed-phase reinforcement in metal alloys to anisotropic dispersed-phase reinforcement in polymers

Tandon and Weng [16] make an important transition with the above arguments for metallurgy to polymer composites. Utilizing Mori and Tanake's derivations, the sum of the applied stress on the composite plus the perturbed stress in the continuous phase from the dispersed phase equals the elastic constants times the sum of the strain in the continuous phase plus the average perturbed strain from the dispersed phase.

Of the five independent elastic constants that are associated with polymer composites, we will focus on the Young's modulus associated with stress in the direction of the alignment of clay particles. We will consider the special case that Eshelby analyzed for the dispersed phase having the morphology of a disk.

The key considerations in the development of the Young's modulus from the analysis of Tandon and Weng are the aspect ratio (A) of the dispersed phase, the volume fraction of the dispersed phase (Φ_f), and the Poisson's ratio (P) of the continuous phase. Returning to the definition of Young's modulus found in Equation (5.1) above, Eshelby's arguments are employed to determine the stress and strain for the composite. For example, the strain in the composite $(\delta L/L)_c$ will be equal to the strain in the continuous phase plus the product of Φ_f times the strain that the dispersed phase $(\delta L/L)^*$ would experience without being present in the continuous phase. This strain for the dispersed phase is:

$$\left\{K_1(\delta L/L)_1 - K_2\left[(\delta L/L)_2 + (\delta L/L)_3\right]\right\}/K, \tag{5.7}$$

where $(\delta L/L)_1$ is the strain for the continuous phase without the dispersed phase in the direction of the stress, $(\delta L/L)_2$ is the strain for the continuous phase without the dispersed phase perpendicular to the direction of applied stress in the same plane, and $(\delta L/L)_3$ is the strain for the continuous phase orthogonal to the direction of applied stress and $(\delta L/L)_2$. The value for $(\delta L/L)_1$ comes from the Young's modulus of the pure continuous phase (Y_p); $(\delta L/L)_2$ will equal $(\delta L/L)_3$ and can be calculated from the Poisson's ratio (2) and Young's modulus (1):

$$(\delta L/L)_2 = (\delta L/L)_3 = -P(F/\text{Area})/Y_p. \tag{5.8}$$

K_1, K_2, and K were determined by Tandon and Weng from the analysis of the arguments provided by Eshelby.

Returning to the expression for Young's modulus (1) for the composite, one can now express Y_c in relation to Y_p and the derived strain for the composite $(\delta L/L)_c$ as:

$$Y_c/Y_p = \{1 + [\Phi_f(K_1 + 2PK_2)]/K\}^{-1}. \quad (5.9)$$

Eshelby began his analysis by introducing the first (λ) and second (μ) Lamé constants. These constants are employed in describing the relationships between stress, the change in volume of the composite, and the strain in the composite:

$$F/\text{Area} = \lambda(\text{the change in volume/original volume}) + 2\mu(\delta L/L). \quad (5.10)$$

Hence, the stress on the composite is a function of the change in volume [λ(change in volume/original volume)] and the elastic nature of the composite [$2\mu(\delta L/L]$. The Lamé constants are integrated into the calculations of K_1, K_2, and K.

A key relationship in Eshelby's arguments that relate to the calculation of K, K_1, and K_2 is:

$$(\delta L/L)_{1,1} = S_{1,1,1,1}(\delta L/L)^*_{1,1}. \quad (5.11)$$

The example above relates the strain in the composite to the strain in the dispersed phase. For $S_{1,1,1,1}$, the first two subscripts relate to the coordinates for the strain in the composite and the second two subscripts relate to coordinates for the strain in the dispersed phase.

Eshelby used an orthorhombic crystal as an example to describe the spatial relationships of S. He also pointed out that they do not couple ($S_{1,1,2,2} \neq S_{2,2,1,1}$). Mori–Tanaka and Tandon–Weng refer to S as Eshelby's tensors or Eshelby's transformation tensors. Eshelby utilized direction cosines from an "observation point" to a volume element to evaluate the elastic fields in the dispersed phase. $S_{1,1,1,1}$, $S_{2,2,2,2}$, $S_{2,2,3,3}$, $S_{2,2,1,1}$, and $S_{1,1,2,2}$ are needed to calculate K, K_1, and K_2 for Equation (5.9). Their values for disk-shaped dispersed phases are provided in the Tandon–Weng and Eshelby publications. These values are solely a function of aspect ratio and Poisson's ratio. Knowing the proper S values, one can return to Equation (5.11) to calculate the strain in the dispersed phase as a function of the strain in the composite.

One should be cautious when interpreting the results in Tandon and Weng's article in relation to subsequent discussions in this chapter on montmorillonite reinforcement of polymers. Tandon and Weng define the aspect ratio for reinforcing disks in the polymer continuous phase as the thickness of the disk divided by the diameter of the disk. They also align the disks in the polymer so that the thickness of the disk is aligned with the direction of the stress for Young's modulus. Subsequent discussions in this chapter on the montmorillonite reinforcement of polymers will employ a definition of aspect ratio that is inverse to that in the Tandon and Weng article. The montmorillonite particles will be evaluated with the alignment of their major axis in the direction of the applied stress. This distinction is clarified in the discussion at the end of the article.

Professor L. A. Utracki pointed out that the plot provided in the article of Young's modulus versus the volume fraction of the dispersed phase for glass flakes or mica (a small-aspect ratio based on Tandon and Weng's definition) indicates that spherical dispersed glass phases will have a greater Young's modulus than the flakes or mica (this is clearly not the case experimentally). Professor G. J. Weng pointed out in his reply that the mica flakes in his discussion are aligned perpendicular to the direction of stress for the determination of Young's modulus. He also pointed out that when the flakes are aligned with their major axis parallel to the direction of stress, the Young's modulus will be higher than reinforcing spheres and composites reinforced with continuous fibers.

Tandon and Weng compared Equation (5.9) with the Halpin–Tsai Equation (5.6) at three different aspect ratios (100, 10, and 2) as a function of the volume fraction of the dispersed phase (0–0.5). Good agreement between the two equations was observed except at the aspect ratio of 2 for the larger volume fractions.

Equation (5.9) was also compared to the experimental results of Terry Richard [17] on solid glass micropheres reinforcing polyester. The diameter of the glass spheres as 210–297 μm. The polyester–glass sphere composites were prepared by curing the polyester composites in glass-aluminum molds at room temperature. The molds were slowly rotated until the composite gelled to ensure that the glass spheres were distributed uniformly in the composites. The agreement is good up to a volume fraction of about 0.5.

5.6 Validation of the morphology of montmorillonite as anisotropic dispersed-phase reinforcement in polymer

Fornes and Paul [5] expressed concern about the suitability of the morphology of montmorillonite being modeled as a disk in Equations (5.9) and (5.6). The alignment of disks in the direction of stress presents different aspect ratios as a function of the distance from the center of the disk.

Rouse *et al.* [18] developed a technique to visualize individual montmorillonite particles. Pyrolytic graphite is attached to the surface of a silicon wafer. An aqueous solution of poly(diallyl dimethylammonium chloride) is utilized to treat the surface of the graphite with a cationic polyelectrolyte. An aqueous dispersion of montmorillonite that has been centrifuged to remove aggregates of montmorillonite particles is poured onto the graphite surface that contains the polyelectrolyte. The montmorillonite particles are stuck to the graphite surface which allows for particle size determination by SEM.

The montmorillonite that was evaluated by Rouse *et al.* [18] is very similar to the montmorillonite that was evaluated by Fornes and Paul [5]. The morphology of the montmorillonite is in the form of irregular sheets (not disk-shaped) and varies in size from 200 to 600 nm. Ploehm and Liu [19] provide a quantitative evaluation of particle size and particle size distribution of similar montmorillonite evaluated by Rouse *et al.* [18] and Fornes and Paul [5] by AFM. The montmorillonite appears to be in a similar morphology of irregular sheets. Evaluation of the particle size and particle size distribution of montmorillonite particles gave a mean thickness of 0.8 to 1.2 nm; the mean aspect ratio was calculated to be 180 ± 91 nm, and the median value was 160 nm.

The montmorillonite–polymer nanocomposite that was evaluated by Fornes and Paul [5] was prepared by melt-blending nylon 6 (Capron B135WP) with an organomontmorillonite (montmorillonite exchanged with bis(hydroxyethyl)-(methyl)-rapeseed quaternary ammonium chloride at 95 meq/100 g of montmorillonite) with a Haake corotating twin-screw extruder. The montmorillonite content in the polymer was experimentally determined to be 0.65, 1.31, 1.91, and 3.3% by volume by weighing the ash that remained from combusting the polymer–montmorillonite nanocomposite in an oven at 900°C for 45 min.

The Young's modulus was determined for these nanocomposites by ASTM D638. The tensile bars were prepared by injection molding the polymer–montmorillonite pellets from the melt-blending process. The Young's modulus for the polymer–montmorillonite nanocomposites were 3.49, 3.92, 4.59, and 5.70 GPa respectively for the above volume % concentrations of montmorillonite in the polymer.

The aspect ratio and particle orientation of the montmorillonite in the polymer nanocomposites were determined by TEM. The exfoliation of the montmorillonite into individual plates and the alignment of the plates in the direction of the applied stress for the determination of the Young's modulus were found to be very good.

The determination of the aspect ratio of the montmorillonite by TEM is fraught with difficulties. Fornes and Paul point out that the utilization of a two-dimensional representation (TEM) of a three-dimensional particle in a polymer matrix would be expected to provide dimensions that are smaller than the actual dimensions. The sections obtained by microtoming for the TEM were on the order of 50 nm thick. This thickness is significantly less than the expected dimensions of the montmorillonite particles which add to the complexity of the particle size determination by TEM.

The authors presented a detailed protocol for the evaluation of particle size which included the preparation of a transparency of the TEM images, digitizing the images, and image analysis (Scion Image 1.63 MacOS, NIH). The number-average particle length was measured to be 91 nm; the weight-average particle length was measured to be 118 nm. Notice the difference in length between these results and those determined by Rouse *et al.* [18] and Ploehn and Liu [19] above.

The montmorillonite in all three studies came from the same geological deposit and was processed in an identical fashion at Southern Clay Products in Gonzales, Texas. The average number of montmorillonite particles in each dispersed-phase particle was measured to be 1.37; the weight average was measured to be 1.67. The 001d spacing obtained from wide-angle X-ray spectrometry was utilized to determine the average thickness of the dispersed-phase particles. The average number thickness of the particles in the dispersed phase of the polymer was calculated to be 1.61 nm; the weight-average thickness was calculated to be 2.43 nm. Hence, the aspect ratio for the number-average values was calculated to be 57; the aspect ratio for the weight-average numbers was calculated to be 49.

The Young's modulus of the montmorillonite particles is assumed to be similar to mica, 178 GPa. The Young's modulus of the polymer is 2.75 GPa. The Poisson ratio of the polymer is assumed to be 0.35 and 0.20 for the montmorillonite.

With these values, a comparison was made between the predicted values from the Halpin–Tsai and Mori–Tanaka models and the measured Young's modulus as a function of montmorillonite concentration in the polymer. When the experimentally determined number-average aspect ratio, 57, was employed, Halpin–Tsai predicted higher values and Mori–Tanaka predicted lower values than the experimental results. When the aspect ratio for the theoretical perfect exfoliation of montmorillonite was utilized in the Mori–Tanaka model, 97, the values were virtually identical to the experimental values. When the Halpin–Tsai model was altered to accommodate dispersed phases that have more than one montmorillonite particle with a number-average length of 91 nm and a spacing between each montmorillonite particle in the dispersed phase of 1.8 nm, the experimental values correspond to a value of 1.4 montmorillonite plates for each dispersed-phase particle. This is very close to the experimentally determined number-average value of 1.37. When the same substitution is made in the Mori–Tanaka model, the experimental values line up with the predicted values for full exfoliation.

This examination of experimentally determined Young's modulus values for montmorillonite–nylon 6 nanocomposites compared with the Halpin–Tsai and Mori–Tanaka models seems to indicate that the extremely effective reinforcing efficiency of montmorillonite in polymer can be explained by its high modulus and large aspect ratio when fully exfoliated.

5.7 Refinement of the mechanism of montmorillonite reinforcement of polymers

Concern about the applicability of a disk shape as a model for montmorillonite reinforcement for the above evaluations of organomontmorillonite exfoliated into nylon 6 encouraged Lee and Paul [20] to evaluate an ellipse-shaped model as the dispersed phase for montmorillonite. This concern is reinforced by the work of Rouse *et al.* [18].

TEM and scanning transmission electron microscopy (STEM) of the same organomontmorillonite–nylon 6 nanocomposite by Yoon *et al.* [21] seemed to be consistent with the observations by Rouse *et al.* [18]. TEM and STEM photographs of montmorillonite in nylon 6 taken parallel to the normal direction of the injection-molded polymer nanocomposites indicate irregular morphology that would not be duplicated by a regular disk shape.

Lee and Paul [20] utilized similar arguments developed by Eshelby [15], Mori and Tanaka [9], and Tandon and Weng [16] discussed above, for modeling ellipse-shaped dispersed-phase reinforcement. Lee and Paul [20] identify the dimensions of the ellipse as $a_1 > a_2 > a_3$. Eshelby assigned the dimensions of the ellipse as $a > b > c$. Tandon and Weng describe their ellipse dispersed phase as 2 l for the major axis and 2 d for the minor axis. Hence, $b = c$ and $a_2 = a_3$ when comparing Tandon's and Weng's arguments with Eshelby's and Lee's and Paul's arguments. A significant improvement is found in Lee's and Paul's discussion when compared to the discussions of Eshelby and Tandon and Weng. Lee and Paul provide a flow diagram for the calculation of modulus from the same arguments found in Eshelby and Tandon and Weng. This provides a much clearer picture of the sequence for the calculation procedures.

Equation (5.9) was evaluated by Lee and Paul with the ellipse morphology in relation to the disk morphology that was evaluated by Fornes and Paul for the organomontmorillonite–nylon 6 evaluations discussed above.

The Halpin–Tsai model was also compared with the ellipse model predictions. The modulus for montmorillonite (178 GPa), the Poisson ratio for the nylon 6 (0.35), the Poisson ratio for the montmorillonite (0.20), and the modulus for nylon 6 (2.75 GPa) that were employed in the Fornes–Paul evaluations were utilized in the calculations [5]. The volume fractions for the dispersed phase were similar to the experimental values determined in the Fornes–Paul work. When the major axis (a_1) of the dispersed-phase ellipse is aligned with the direction of the applied stress and the ratio of a_1/a_2 is greater than 1 (1 is the ratio for a disk morphology), Y_c/Y_p is greater for the ellipse when compared to the disk morphology. The application of the ellipse morphology in place of the disk morphology allows for predicted values for Y_c/Y_p to be closer to the experimental results provided in the work of Fornes and Paul.

Plots are provided in the Lee–Paul discussion of Y_c/Y_p as a function of a_1/a_2, volume fraction of dispersed phase, and a_1/a_3. The sensitivity of Y_c/Y_p to changes in the ellipse morphology indicates that the ellipse provides a more flexible model than the disk model for adaptation of a model to experimental results.

Work by Wang and Pyrz [22,23] affirms the work above. Their work begins with the Eshelby thesis described above and focuses on the development of the Mori–Tanaka approach to predict the modulus of the composites. The two morphologies that are examined are oblate spheroids (flattened spheres with the limit being a plate) and prolate spheroids (stretching or elongating a sphere with the limit being a fiber).

One should be cautious in reviewing this work. The definition of the aspect ratio for the oblate spheroids (the thickness divided by the long dimension) is the inverse of what is described in the bulk of this chapter. There is a difference in assigning the Young's modulus in the first article (178 GPa) and the second article (167 GPa) for montmorillonite. 178 GPa is generally accepted as the Young's modulus for montmorillonite in the bulk of the work featured in this chapter. The bulk and the shear moduli are calculated from the model and utilized to calculate the Young's modulus.

An unusual feature of the work is the modeling of an isotropic (random) distribution of the montmorillonite within the polymer (continuous) matrix. The work above focused on modeling the composites with the montmorillonite aligned in the direction of the applied stress. This alignment is important for the full benefit of reinforcements of the anisotropic morphology of montmorillonite to be realized. This isotropic model predicts the modulus for epoxy and high viscosity polyester–montmorillonite composites well.

The difficulty of aligning montmorillonite in these high-viscosity polymers is discussed later in this chapter. This alignment difficulty as a function of viscosity results in a random distribution of the montmorillonite. An additional feature of this work is a prediction of the benefit of the constrained polymer region in relation to the surface of the montmorillonite. A measurement on the constrained polymer region is found in the chapter on barrier properties.

Hbaieb *et al.* [24] compared the utility of modeling polymer–montmorillonite nanocomposites by finite element analysis (FEA) in relation to the Mori–Tanaka model. The three-dimensional finite element model (FEM) was found to be superior to the two-dimensional one. For the calculations, the aspect ratio (A) was chosen to be 50, $Y_f/Y_p = 100$, the Poisson ratio for the polymer (P_p) was assumed to be 0.35, and the Poisson ratio for the montmorillonite was assumed to be $P_f = 0.2$. The FEA was performed using the commercial package, ABAQUS. The morphology of the montmorillonite was assumed to be disk-shaped. The limitations of this assumption are exposed in the discussion above by Lee and Paul.

The size of the volume element for the FEA is critical. The desire to minimize the size of the volume element is understandable as regards minimizing the computation time. The volume elements were chosen to contain 50–100 particles. Symmetry boundary conditions were employed because of the sectioning of particles that partially lie outside the volume element. Two different configurations were considered; particles aligned in the direction of applied stress and a random orientation of particles in relation to the direction of applied stress.

The Mori–Tanaka and the FEA tracking agree up to approximately 5% loading of the montmorillonite. Two interesting results emerge at higher loading levels when the Mori–Tanaka model is compared with the FEA. For oriented particles in the direction of the applied stress, the FEA predicts greater reinforcement. Presumably, the FEA predicts that particle–particle interactions (as a result of closer spacing between the particles from crowding at higher concentrations) provide for a higher overall aspect ratio and hence, greater reinforcement. With random particle orientation, the Mori–Tanaka model predicts a higher reinforcement. Particularly at higher concentrations, FEA predicts that particles will align with one another in domains, resulting in an overall lower aspect ratio.

This lower aspect ratio from self-alignment in the random orientation will result in lower reinforcement. We will explore this possibility when epoxy-montmorillonite nanocomposites are discussed later in chapter 6. Because of the high viscosity of epoxies, it is difficult to align montmorillonite in the polymer matrix. Alignment is possible with the application of a magnetic field to influence the orientation of montmorillonite through their inherent magnetic properties.

The limitations of the two-dimensional FEM model compared with the three-dimensional model are illustrated in the work of Sheng *et al.* [25]. Caution is advised for interpreting the analysis of this work. The chemical structure provided in Table 2 on page 496 of reference [25] is not correct for montmorillonite. The chemical structure that is provided is for pyrophyllite (a clay without inherent charge). The unit cell composition of sodium counterion montmorillonite with a 91.5 CEC is $[(Si_8)(Al_{3.33}Mg_{0.67})\ O_{20}(OH)_4]Na_{0.67}$. Hence, the molecular weight is also incorrect. Fortunately, the error is not dramatic enough to seriously compromise the general arguments that are developed.

The polymer–montmorillonite composite examples that were utilized to test the model in this work were not ideal. The TEM of the montmorillonite in the epoxy composite pictured on page 493 were not aligned. The TEM of the MXD6–montmorillonite composites pictured on page 494 are poorly exfoliated and not aligned at the higher montmorillonite concentration. WAXS was described ((001) $d = 4.1$ nm) but the spectra were not provided. The organomontmorillonite that was employed to make the MXD6–montmorillonite composite was I.34MN from Nanocor Inc. The composition of this organomontmorillonite is similar to Cloisite 30B. The montmorillonite concentrations in the MXD6 were 1.1, 3.67, 4.17, and 5.27 wt. %. The test samples were injection-molded. No processing details were provided. The Mori–Tanaka model and the FEM model in this work provide a good prediction of the experimental results of the MXD6–montmorillonite composites when one considers the poor exfoliation of the montmorillonite in the MXD6 and the inherent error associating the wrong formula for montmorillonite in this work.

A much-improved prediction of modulus is obtained with the FEM model when data from Fornes *et al.* [26] was utilized from nylon 6–montmorillonite composites that had improved exfoliation.

A critical factor in the successful utilization of the FEM model to predict the mechanical properties of polymer–montmorillonite nanocomposites is the adoption of the proper representative volume element (RVE) for the composite. The work of Tsai and Sun [27] illustrates this importance. They employ a two-dimensional model that explores the relationship of fully exfoliated and intercalated composites for successful FEM analysis. The load transfer efficiency in relation to the RVEs as regards the stress distribution between the montmorillonite and the polymer matrix was calculated utilizing the shear lag model. The particles are ideally aligned in the direction of the applied stress. The distance between the unit cell of the composite and the ends of the particles is not considered in the load transfer analysis. The FEM analysis and this shear lag model correspond well in predicting the effective length of reinforcement of the montmorillonite when the aspect ratio of the montmorillonite is equal to or greater than 400 for fully exfoliated particles that are ideally aligned in the applied stress direction. As in reference [27], the prediction of the effective length of reinforcement for montmorillonite particles in intercalated composites becomes more problematical as the number of montmorillonite particles increases in each intercalated domain.

Full exfoliation and alignment of the montmorillonite particles in polymers provides the maximum reinforcement potential in polymer–montmorillonite nanocomposites. Evaluation of the morphology of montmorillonite in polymers is usually provided by WAXS and a TEM of the polymer composite.

Schaefer and Justice [28] suggest an additional parameter that would be missed by WAXS and TEM. They suggest that large-scale particle–particle associations and particle randomization of fully exfoliated polymer–montmorillonite nanocomposites compromise the ultimate mechanical performance. For fully exfoliated nanocomposites, ultra-small-angle X-ray scattering (USAXS) is needed to identify the presence of large-scale particle associations.

TEM must also be employed to distinguish between scattering from polymer coils and large-scale particle associations. One example cited by Schaefer and Justice is work done by Fornes and Paul [29] that is an extension of their above work [5] with nylon 6–montmorillonite nanocomposites. They evaluated the role of extending the alkyl chain length of nylon (nylon 11 and 12) in exfoliation efficiency when compared to nylon 6. They conclude that the increased hydrophobic nature of nylon 11 and 12 when compared with nylon 6 is responsible for the compromise of exfoliation efficiency of the organomontmorillonite in nylon 11 and 12. The change in mechanical properties of the nylon 11 and 12 is also the result of their increased ductility when compared to nylon 6. Schaefer and Justice plotted the modulus results from nylon 6, 11, and 12 versus the volume fraction of montmorillonite and compared the predicted aspect ratio with the USAXS data [30] of montmorillonite in water that indicated large-scale ordering of the montmorillonite. The aspect ratio derived from USAXS predicted an underestimation of the mechanical performance measured by Fornes and Paul for nylon 6, 11, and 12. Schaefer and Justice suggest that this difference is the result of the large-scale associations of montmorillonite that would not be detected by WAXS and TEM.

Schaefer *et al.* [30] provided additional supporting data for long large-scale ordering with the exfoliation of organomontmorillonite in nylon 6,6. Cloisite Na, utilized to generate the USAXS data discussed above, was modified with a silane (tert-butyldimethylchlorosilane) before it was exchanged with dioctadecyl dimethyl quaternary ammonium chloride. The nylon 6,6 had a $M_n = 1.67 \times 10^4$. A twin-screw extruder was employed to prepare the polymer nanocomposite. USAXS data seems to indicate disk morphology for the montmorillonite up to approximately 10 nm in size. The data implies large-scale disorder for particles greater than 10 nm.

A model was introduced that describes the montmorillonite as flexible. This flexibility could result in various configurations other than flat sheets, particularly at dimensions larger than 16 nm. More USAXS work of polymer–montmorillonite nanocomposites needs to be performed before any definite relationships can be established for long-range order in polymer–montmorillonite composites.

5.8 Conclusions

The Halpin–Tsai and Mori–Tanaka theories based on first-principle arguments adequately model the mechanical properties provided by the reinforcement of montmorillonite as a dispersed phase in all polymers [5,20]. The significant independent variables that correlate to reinforcement are: aspect ratio, modulus, and the alignment of the montmorillonite in the direction of the applied stress. It is enlightening to examine some of the general predictions of these theories as they relate to the effect of clay loading, aspect ratio, and the modulus of the pristine

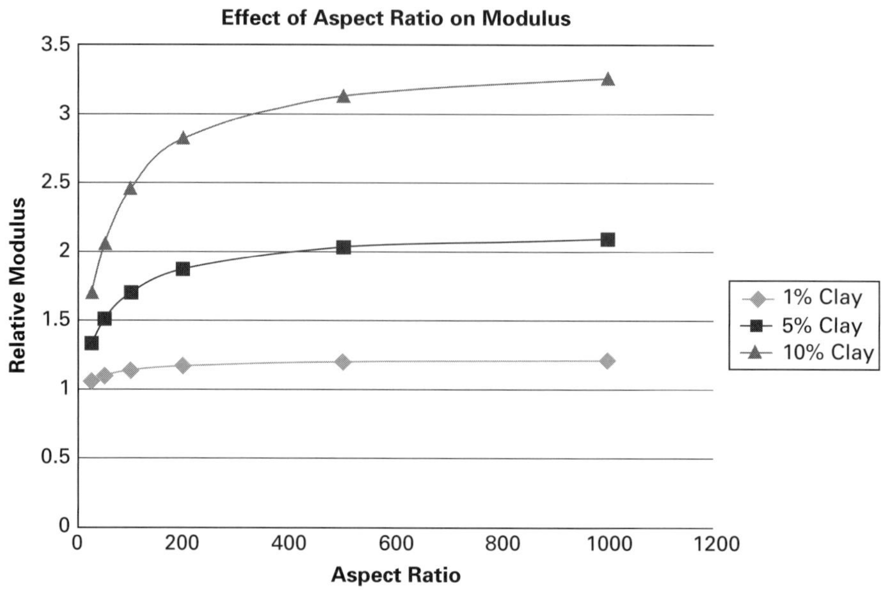

Figure 5.1 Effect of aspect ratio on relative modulus increase.

polymer. In summary, only predictions using the Halpin–Tsai equation will be presented below. The exact same trends are seen with the Mori–Tanaka theory.

The first prediction is the effect of the aspect ratio on the relative modulus. Figure 5.1 demonstrates that the increase in modulus starts to plateau in the range of an aspect ratio of 200 and only a modest increase occurs for very large aspect ratios. At an aspect ratio of 200, a relative modulus increase of 2.8 is observed, while at an aspect ratio of 1000, the increase is 3.25. This is a gain of 16% in relative modulus for a five fold increase in aspect ratio. This is fortuitous, since montmorillonite's aspect ratio is in the range of 200 for many natural deposits. It would therefore appear to be counterproductive to make composites with very high aspect ratio materials.

The second variable of interest is the effect of clay loading on the relative modulus. In the case of macro and microscale fillers, there is a linear increase in modulus with loading levels as high as 40–50%. This is not the case with nanoclays, as can be seen in Figure 5.2. This linear increase in modulus only holds for loadings below 7%. Above that point the curve begins to plateau. This trend has been observed in many composites and indicates that in most nanocomposites, loadings in the range of 5 to 7% are optimal.

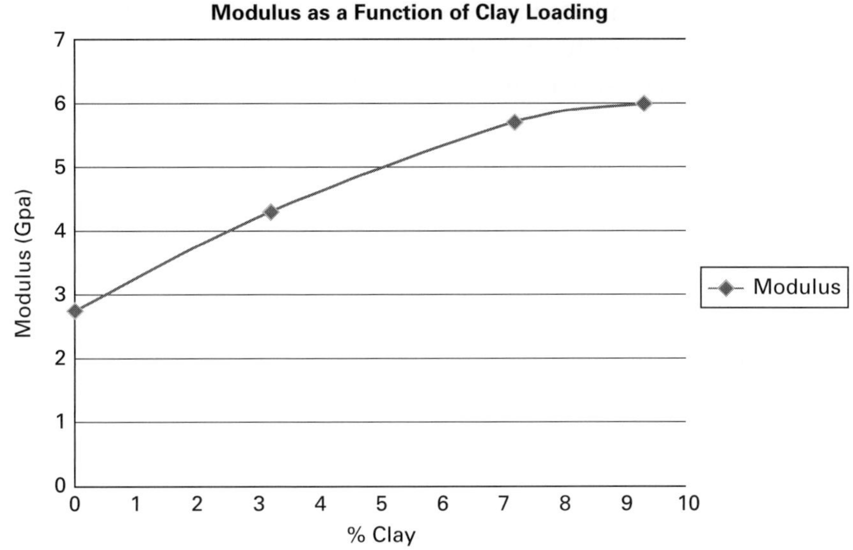

Figure 5.2 Effect of clay loading on modulus increase.

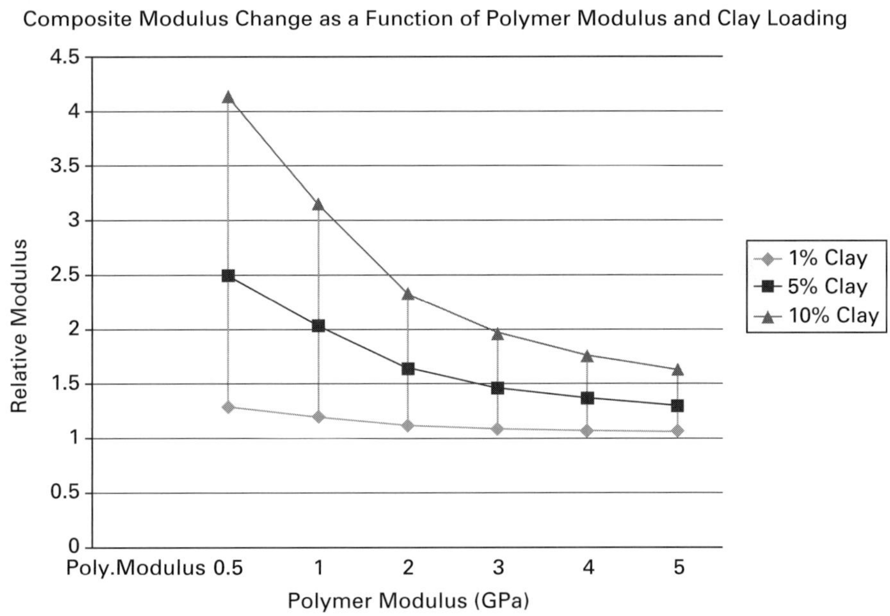

Figure 5.3 Effect of polymer modulus on relative modulus increase at three different clay loadings.

The final variable of interest is the effect on relative modulus for different polymers used in the composite. In Figure 5.3, the change in relative modulus as a function of the starting modulus of the pure polymer is plotted at three different clay loadings. It can be seen that the relative modulus increase for a low-modulus polymer is far greater than for a very stiff high-modulus polymer. For example, a nanocomposite made with 5% loading of clay in a polymer with a modulus of 0.3 GPa would have a 250% increase in modulus. The same loading in a polymer with a starting modulus of 3.0 GPa would show an increase of 50%.

References

[1] G. H. Michler and F. J. Balta-Calleja (eds). *Mechanical Properties of Polymers based on Nanostructure and Morphology* (CRC Press Taylor & Frances Group, Boca Raton, FL, 2005).

[2] Y.-W. Mai and Z.-Z. Yu, eds. *Polymer Nanocomposites* (CRC Press LLC, Boca Raton, FL, 2006).

[3] L. A. Utracki. *Clay-containing Polymer Nanocomposites, vols 1 and 2* (Rapra Technology Limited, Shawbury, UK).

[4] H. Van Olphen. *An Introduction to Clay Colloid Chemistry for Clay Technologists, Geologists and Soil Scientists* (Interscience Publishers, New York, 1963).

[5] T. D. Fornes and D. R. Paul. Modeling properties of nylon 6/clay nanoparticles using composite theories. *Polymer*, **44** (2003), 4993–5013.

[6] J. C. Halpin. Stiffness and expansion estimates for oriented short fiber composites. *Journal of Composite Materials*, **3** (1969), 732–734.

[7] J. C. Halpin and J. L. Kardos. The Halpin–Tsai equations: a review. *Polymer Engineering and Science*, **16** (5) (1976), 344–352.

[8] J. E. Ashton, J. C. Halpin, and P. H. Petit. *Primer on Composite Materials* (Analysis Techomic Publishing Co., Stamford, Conn., 1969).

[9] T. Mori and K. Tanaka. Average stress in matrix and average elastic energy of materials with misfitting inclusions. *Acta Metallurgica*, **21** (1973), 571–574.

[10] R. Hill. Elastic properties of reinforced solids: some theoretical principles. *Journal of the Mechanics and Physics of Solids*, **11** (1963), 357–372.

[11] R. Hill. Theory of mechanical properties of fibre-strengthened materials: I elastic behaviour. *Journal of the Mechanics and Physics of Solids*, **12** (1964), 199–212.

[12] R. Hill. Theory of mechanical properties of fibre-strengthened materials: II inelastic behaviour. *Journal of the Mechanics and Physics of Solids*, **12** (1964), 213–218.

[13] J. J. Hermans. The elastic properties of fiber reinforced materials when the fibers are aligned. *Proceedings of the Koninklijke Nederlandse Akademie van Wetenschappen. Series B. Physical Sciences*, **70** (1967), 1–9.

[14] E. H. Kerner. The elastic and thermo-elastic properties of composite media. *Proceedings of the Physical Society*, **69** (1956), 808–813.

[15] J. D. Eshelby. The determination of the elastic field of an ellipsoidal inclusion, and related problems. *Proceedings of the Royal Society of London. Series A, Mathematical and Physical Sciences*, **241**:1226 (1957), 376–396.

[16] G. P. Tandon and G. J. Weng. The effect of aspect ratio of inclusions of the elastic properties of unidirectionally aligned composites. *Polymer Composites*, **5**:4 (1984), 327–333.
[17] T. G. Richard. The mechanical behavior of a solid microsphere filled composite. *Journal of Composite Materials*, **9** (1975), 108–113.
[18] J. H. Rouse, S. T. White, and G. S. Ferguson. A method for imaging single clay platelets by scanning electron microscopy. *Scanning*, **26** (2004), 131–134.
[19] H. J. Ploehn and C. Liu. Quantitative analysis of montmorillonite platelet size by atomic force microscopy. *Industrial and Engineering Chemistry Research*, **45**:21 (2006), 7025–7034.
[20] K. Y. Lee and D. R. Paul. A model for composites containing three-dimensional ellipsoidal inclusions. *Polymer*, **46** (2005), 9064–9080.
[21] P. J. Yoon, T. D. Fornes, and D. R. Paul. Thermal expansion behavior of nylon 6 nanocomposites. *Polymer*, **43** (2002), 6727–6741.
[22] J. Wang and R. Pyrz. Prediction of the overall moduli of layered silicate-reinforced nanocomposites – part I: basic theory and formulas. *Compositer Science and Technology*, **64**:7–8 (2004), 925–934.
[23] J. Wang and R. Pyrz. Prediction of the overall moduli of layered silicate-reinforced nanocomposites – part II. Analyses. *Compositer Science and Technology*, **64** (2004), 935–944.
[24] K. Hbaieb, Q. X. Wang, Y. H. J. Chia, and B. Cotterell. Modelling stiffness of polymer–clay nanocomposites. *Polymer*, **48** (2007), 901–909.
[25] N. Sheng, M. C. Boyce, D. M. Parks, G. C. Rutledge, J. I. Abes, and R. E. Cohen. Multiscale micromechanical modeling of polymer–clay nanocomposites and the effective clay particle. *Polymer*, **45** (2004), 487–506.
[26] T. D. Fornes, P. J. Yoon, H. Keskkula, and D. R. Paul. Nylon 6 nanocomposites: the effect of matrix molecular weight. *Polymer*, **42** (2001), 9929–9940.
[27] J. Tsai and C. T. Sun. Effect of platelet dispersion on the load transfer efficiency in nanoclay composites. *Journal of Composite Materials*, **38**:7 (2004), 567–579.
[28] D. W. Schaefer and R. S. Justice. How nano are nanocomposites? *Macromolecules*, **40**:24 (2007), 8501–8517.
[29] T. D. Fornes and D. R. Paul. Structure and properties of nanocomposites based on nylon-11 and -12 compared with those based on nylon-6. *Macromolecules*, **37**:20 (2004), 7698–7709.
[30] D. W. Schaefer, R. S. Justice, H. Koerner, R. Vaia, and C. Zhao. Large-scale morphology of dispersed layered silicates. *Materials Research Society Symposium Proceedings*, **840** (2005), 57–62.

6 Variables associated with polymer–clay processing in relation to reinforcement theory

6.1 The polymer as a significant independent variable in the mechanical performance of polymer–clay nanocomposites

Chavarria and Paul [1] performed a complete evaluation of a comparison of the significant variables that relate to the successful exfoliation of organomontmorillonite in nylon 6 with the utility of these variables in the preparation of organomontmorillonite–nylon 6,6 polymer nanocomposites. The equipment and protocol for these evaluations were identical to those found in reference [1] with nylon 11 and 12. The same nylon 6 was evaluated (B135WP). The molecular weight of the nylon 6 was measured to be $M_n = 29\,300$ by intrinsic viscosity. This is slightly different from the reported viscosity, $M_n = 31\,100$, in reference [1]. The same organomontmorillonite (montmorillonite exchanged with octadecytrimethyl quaternary ammonium ion at 95 meq/100 g of montmorillonite) was employed in both studies. The nylon 6,6 was produced by DuPont, Zytel 42A. The molecular weight was not reported.

The production of nylon 6,6 is significantly different from the ring opening polymerization of ε-caprolactam to produce nylon 6. Hexanedioic acid (adipic acid) is neutralized with hexamethylenediamine in a 50% aqueous solution. The pH is carefully monitored in order to ensure the proper stoichiometry of dicarboxylic acid and diamine. The water is boiled away to produce nylon 6,6. One might intuitively anticipate similar properties between nylon 6 and nylon 6,6. This is not the case. The melting point of nylon 6,6, for example, is much higher (262 °C) than that of nylon 6 (219 °C). The two extra methylene units in the diamine between the amide linkages for nylon 6,6 plus the mixture of end groups (some of the polymer chains have only amine end groups; some of the polymer chains have only carboxylic acid end groups) provide significantly different properties for nylon 6,6 when compared to nylon 6. These differences transfer to differences in exfoliation efficiency for nylon 6,6 when compared to nylon 6.

The imaging technique developed for the above work was compared with the use of Adobe Photoshop software to measure particle size and particle size distribution of the montmorillonite in the polymer composite. The Adobe Photoshop method calculated a smaller average length for the particles (more small particles were imaged by the Adobe Photoshop method). The calculated aspect ratios of both methods produced similar results and were commensurate with the experimental data and theoretical predictions. The results indicated that the efficiency in exfoliation of the organomontorillonite in nylon 6 was superior to that in nylon 6,6. The modulus values of the nylon 6,6–montmorillonite polymer nanocomposites were consistently inferior to those of nylon 6. The modulus values for nylon 6,6 were further compromised as a function of increased montmorillonite content in a nonlinear fashion.

Caution is suggested when interpreting the results in Figure 18 for nylon 6,6. The Halpin–Tsai and Mori–Tanaka theoretical value plots seemed to be mislabeled. The dashed plot in Halpin and Tsai should probably be labeled with an aspect ratio of 28 and the solid line for the Mori–Tanaka plot should probably be labeled with an aspect ratio of 65.

The crystalline nature of nylon 6–montmorillonite nanocomposite was established by Kojima *et al.* [2] by X-ray diffraction. The crystalline form is γ and is oriented planar to the direction of the montmorillonite, which is oriented in the direction of the extruded film. Yoo and Paul [3] evaluated the significance of an amorphous polyamide similar to nylon 6 as regards exfoliation efficiency and mechanical properties. The amorphous polyamide, Zytel 330 manufactured by DuPont, was prepared with the same monomer that is utilized to prepare nylon 6 and a co-monomer, phthalic acid. Note that the molecular weight of the Zytel 330 is $M_n = 14\,000$, and that the nylon 6 study above by Fornes and Paul was when $M_n = 29\,300$. Also, because phthalic acid is the co-monomer with 6-aminohexanoic acid to produce the amorphous polyamide, the number of carboxylic acid end groups on the polymer chains increases significantly to a ratio of 4:5 carboxylic groups to amine groups. Nylon 6 is prepared commercially from ε-caprolactam. Hence, the ratio of carboxylic acid groups to amine groups at the end of the polymer chains is approximately 1.

The role of the hydrophobic surface on the organomontmorillonite was evaluated by utilizing different quats that were ion exchanged onto the surface of the montmorillonite. Bis(2-hydroxy-ethyl)methyl tallow quaternary ammonium-exchanged montmorillonite (Cloisite 30B) would probably be considered to be the most hydrophilic organomontmorillonite, as a result of the influence of the two hydroxyl functionalities. Trimethyl hydrogenated tallow quaternary ammonium ion-exchanged montmorillonite would be considered to be more hydrophobic. The most hydrophobic organomontmorillonite of the series utilized dimethyl bis(hydrogenated tallow) quaternary ammonium ions (Cloisite 20A) for the surface treatment.

The exact same process with the same equipment that was employed to produce the nylon 6 nanocomposites was employed with the amorphous polyamide. The same procedure as outlined above with the nylon 6 nanocomposites was used to characterize the amorphous polyamide nanocomposites. Cloisite 30B and the organomontmorillonite prepared with trimethyl hydrogenated tallow quat demonstrated excellent exfoliation. Cloisite 20A was not as effective in exfoliation. The mechanical performance of the montmorillonite exchanged with trimethyl hydrogenated tallow quat had the superior Young's modulus.

The Young's modulus for the nylon 6 and the amorphous nylon are very similar: about 2.75 GPa. When one compares the change in modulus as a function of wt.% of montmorillonite in the two nylons, the Young's modulus values are almost identical. The lower molecular weight amorphous nylon with the higher aromatic content and higher concentration of carboxylic functionality on the ends of the polymer seems to prefer a slightly more hydrophobic surface treatment on the montmorillonite for excellent exfoliation and mechanical performance when compared to the nylon 6 in the above study.

Fornes *et al.* [4] explored the role of different nylon 6 molecular weights on the mechanical performance of nylon 6–montmorillonite nanocomposites. A higher molecular weight nylon 6 ($M_n = 29\,300$) and a lower molecular weight nylon 6 ($M_n = 16\,400$; note the similarity to the molecular weight of the amorphous nylon evaluated above) than the nylon 6 evaluated above were processed into polymer nanocomposites with the same equipment, organomontmorillonite, and protocol utilized above to prepare nylon 6–montmorillonite nanocomposites.

Evaluations of the exfoliation efficiency of the montmorillonite in the polymers and mechanical performance were identical to the protocol for evaluating the nylon 6–montmorillonite nanocomposite above. Data for the weight concentrations of montmorillonite in the three nylon 6 polymers were at approximately 3% and 7% and were reported in tabular form. Plots of mechanical performance versus wt.% montmorillonite concentration indicated that montmorillonite concentrations at approximately 1.5%, 4%, and 6.5% were also evaluated.

Exfoliation of the montmorillonite in the lower molecular weight nylon 6 was compromised. The exfoliation of montmorillonite into the higher molecular weight nylon 6 was excellent and comparable to the exfoliation in the nylon 6 evaluated above. The polymer nanocomposite that was prepared with the high molecular weight nylon 6 demonstrated superior modulus, percent elongation to break, and Izod impact strength when compared to the performance of the nylon 6 nanocomposites at the lower molecular weights. However, the Young's modulus values with the high molecular weight nylon 6 were not significantly different from the values reported in the work above with nylon 6 nanocomposites.

An argument for the difference in mechanical performance as a function of molecular weight of the nylon 6 was predicated on the lower viscosities of the lower molecular weight nylon 6 polymers compromising the efficiency of the processing protocol necessary to produce full exfoliation of the organomontmorillonite. An experiment with a higher molecular weight amorphous nylon 6 could prove to be a valuable addition to the work above.

6.2 Processing as a significant independent variable for polymer–clay nanocomposite preparation

All of the above work was carried out with the same processing equipment and conditions. To explore the significance of different processing equipment in regard to mechanical performance, Dennis *et al.* [5] employed four different extruders to prepare nylon 6 montmorillonite nanocomposites. The nylon 6 for this study is the higher molecular weight version that was evaluated above. The two organomontmorillonites that were evaluated were Cloisite 30B and Cloisite 15A.

The choice of Cloisite 30B has been demonstrated in the work above as an excellent choice for exfoliation into nylon 6. The choice of Cloisite 15A is not good, based on the work above. One must be cautious when comparing the results with Cloisite 15A and Cloisite 20A. Both organomontmorillonites utilize the same quat treatment. However, Cloisite 15A has an excess of approximately 30 meq/100 g of montmorillonite of quat past the exchange capacity of the montmorillonite. This excess quat could diffuse into the polymer and alter the mechanical properties of the composite. The counterion of this excess quat will be chloride. The montmorillonite content in the polymer was 3.7 wt.% for Cloisite 30B and 3.1 wt.% for Cloisite 15A.

The four different extruders were:

- a Killion 25.4 mm single-screw extruder with a high intensity mixing head;
- Japan Steel Works 30 mm modular self-wiping corotating twin screws;
- Leistritz 34 mm with modular intermeshing counter-rotating twin screws;
- Leistritz 34 mm with modular nonintermeshing counter-rotating twin screws.

The conditions of low, medium, and high shear in the extruders were a function of the screw configurations. Mean residence time (47–162 s) in the extruders was the additional significant independent variable. The degree of exfoliation was determined by the methods discussed above. Unfortunately, Cloisite 30B was evaluated at only two sets of conditions (single-screw and counter-rotating intermeshing screws at medium shear conditions). Exfoliation was very good with the twin-screw conditions and poor with the single-screw configuration. Cloisite 15A was evaluated at all the processing conditions, and achieved good exfoliation with the counter-rotating intermeshing twin-screw setup with medium shear conditions. High shear conditions seem to compromise the exfoliation efficiency of Cloisite 15A.

Nonequilibrium thermodynamic arguments by Rubi [6] and Onsager's reciprocity relationship could be applicable to the exfoliation behavior of Cloisite 15A as a function of processing conditions. The conditions that produced the exfoliated state for Cloisite 15A could represent a nonequilibrium thermodynamic state with the nonexfoliated states representing conditions closer to equilibrium. The implications of nonequilibrium thermodynamics to exfoliation of montmorillonite in polymers will not be pursued because those arguments lie outside the scope of this chapter.

The Young's modulus for the exfoliated Cloisite 30B composite was measured at 4.4 GPa; the Young's modulus for exfoliated Cloisite 15A was measured at 4.0 GPa. The lower value for the latter compared to the former could be due to the function of the lower montmorillonite concentration in the polymer, the possible lower degree of exfoliation, and the presence of excess quat.

A more pertinent evaluation of the roll of processing conditions to exfoliation of montmorillonite in nylon 6 to the discussions above is provided by Chavarria *et al.* [7]. The nylon 6 that was evaluated was the higher molecular weight version that was evaluated above. The organo-montmorillonite was prepared by exchanging trimethyl hydrogenated tallow quaternary ammonium ion onto the montmorillonite. This is the same organomontmorillonite that was evaluated above with the amorphous nylon and demonstrated excellent exfoliation efficiency.

Three different extruders were evaluated in this work. One of the extruders was the Haake twin screw (30.5 mm diameter and a length: diameter ratio of 1:10) that was employed in much of the above nylon 6 organomontmorillonite exfoliation studies. The other two were a Werner and Pfleiderer ZSK twin-screw extruder with a diameter of 25 mm and a length: diameter ratio of 48:1 and a microcompounder manufactured by DSM (this is a batch compounder (3 g sample sizes) with a vertical barrel that is fitted with two conical screws that vary in diameter from 0.43 to 1 cm and are 10.75 cm in length).

An important distinction between the processing conditions was in the type of injection molder used to prepare the samples for testing. Polymer nanocomposites from the Werner and Pfleiderer, and Haake extruders were molded with an Arburg Allrounder 305–210–700 injection molder (the same machine that was employed in most of the studies above). Because of the small sample size produced by the microcompounder, the test samples from the microcompounder were prepared with a DSM bench-top microinjection molder.

Two different protocols were employed with the Werner and Pfleiderer extruders to prepare melt-blended polymer nanocomposites. One protocol was identical to that utilized with the Haake; a dry blend of polymer and organomontmorillonite was introduced into the extruder. The other protocol involved introducing the organomontmorillonite into the extruder after the nylon 6 was introduced and melted.

The evaluation of exfoliation efficiency and Young's modulus of the polymer nanocomposites was identical to the procedures described above. The mean residence time was 0.95 min for the Werner and Pfleiderer extruder when the polymer and organomontmorillonite were added separately, 1.38 min for them when the polymer and organomontmorillonite were added together, and 3.4 min for the Haake extruder. The residence time in the DSM extruder varied from 10, 15, and 20 min.

Two different montmorillonite concentrations were evaluated for each process. The low concentrations varied from 0.89% for the DSM to 1.28% for the Haake extruder, and 1.21% to 1.51% wt.% for the Werner and Pfleiderer extruder. The higher concentrations varied from 4.06% for the DSM extruder to 4.37% for the Haake extruder, and from 4.68% to 5.02% for the Werner and Pfleiderer extruder. Exfoliation efficiency was generally good for all three processes.

The Young's modulus was the highest for the polymer nanocomposites prepared with the Werner and Pfleiderer extruders when the organomontmorillonite was added downstream to the molten polymer. The lowest Young's modulus value was obtained when the polymer was added together with the organomontmorillonite. The Haake extruder produced polymer composites with the second highest modulus values, followed by the samples prepared with the DSM extruder.

Increasing the residence time in the DSM extruder from 10 to 20 min had little impact on the modulus values. The Young's modulus steadily increased as a function of the number of passes through the Haake and the Werner and Pfleiderer extruder when the composite was prepared by adding the polymer and organomontmorillonite together.

An examination of the alignment of the montmorillonite particles in the test samples indicated that the greatest alignment occurred when the test samples were prepared with the Arburg injection molder. The DSM injection molder developed less injection pressure and shear environment when compared to the Arburg. These variables for injection molding should be considered when designing molds for polymer–montmorillonite nanocomposites. Proper alignment of the exfoliated montmorillonite particles is key to realizing optimal mechanical performance in a part.

An examination of the particle morphology in the test samples as a function of the flow direction in the injection molder in relation to the transverse direction indicated that the particles are longer in the flow direction when the polymer nanocomposites were prepared with the Werner and Pfleider extruders. This observation is commensurate with the preferred morphology of montmorillonite as an ellipsoid rather than a disk in the modeling studies above. The opposite orientation was observed with test samples prepared from the Haake and the DSM extruders. The difference in the shear elements of the three different extruders could be responsible for the differences in the anisotropic orientations of the montmorillonite particles in the polymer.

6.3 Hydrophilic–hydrophobic balance of the surface of montmorillonite as a significant independent variable for polymer–clay nanocomposite preparation

Fornes *et al.* [8] studied the effectiveness of an extensive array of quaternary ammonium ion surface treatments on montmorillonite in regard to exfoliation efficiency in the higher and lower molecular weight nylon 6 studied above. The Haake extruder was utilized with the same processing protocol as described above. Characterization of the degree of exfoliation and the measurement of the Young's modulus of the polymer nanocomposites were identical to the procedures utilized earlier.

The independent variables for the study were:

- the chain length of the alkyl group of bis(2-hydroxy-ethyl)methyl alkyl quaternary ammonium ions (the alkyl chains varied from an average carbon chain length of 12, 18, and 22);
- tallow versus hydrogenated tallow for trimethyl alkyl quaternary ammonium ions;
- protonated tertiary amine of dimethyl hydrogenated tallow versus the quaternary ammonium ion of trimethyl hydrogenated quaternary ammonium ion (note that the protonated tertiary amine will be a weak acid, unlike the quat);
- methyl dehydrogenated tallow tertiary amine versus the dimethyl dehydrogenated quaternary ammonium ion;
- the level of dimethyl dehydrogenated tallow quaternary ammonium past the exchange capacity of the montmorillonite (Cloisite 20A versus Cloisite 15A).

The concentrations of montmorillonite that were melt-blended into the nylon 6 were generally at two different concentrations for each different organomontmorillonite. The low concentrations were in the range of 3 wt.%. The high concentration was in the range of 4.5 wt.%. The results of this study were consistent with the results reported above.

The additional information that is provided in this study that was not provided above is the chain length of the alkyl group associated with the bis(2-hydroxyethyl)methyl alkyl quaternary ammonium ions appears not to be a significant variable for exfoliation of the organomontmorillonite. The performance of the protonated tertiary amine quats versus the tetra-alkyl quaternary ammonium ion quats exchanged onto montmorillonite seems to offer no significant performance advantage for the single long-chained alkyl quats but seems to provide a performance advantage for the quats with two long alkyl chains.

Xie *et al.* [9] examined the fractured surface of nylon 6–montmorillonite nanocomposites with unexpected results. The organic treatment for montmorillonite that provided exfoliated nanocomposites was bis (2-hydroxyethyl)octadecyl amine. The quat versions of this amine were evaluated above. The montmorillonite had a lower exchange capacity (85 meq/100 g of montmorillonite) when compared to the montmorillonite utilized above. The molecular weight for the nylon 6 (Capron 8200 NL) that was utilized in the study was not provided. It is assumed that the molecular weight is similar to the low molecular weight nylon 6 (Capron 8202) evaluated above [4].

The Young's modulus values were similar to those reported with the low molecular weight nylon. The unusual morphology at the fractured surface examined by SEM for the polymer–montmorillonite nanocomposites at 3% and 5% loading resembled a blooming rosebud. The authors described the morphology as resembling "cabbages." This morphology was not detected with the neat nylon or the unexfoliated polymer nanocomposites.

One could speculate that these unusual structures may be the consequence of potentially long-range structures of the montmorillonite in the polymer that were discussed by Schaefer *et al.* [10,11] earlier. More work is necessary to identify the significant independent variables responsible for this unusual behavior.

Chavarria *et al.* [12] evaluated the relationship between the exchange capacity of the organomontmorillonite and exfoliation efficiency of the organomontmorillonite in nylon 6. Consideration of the erratum [13] is recommended before this work is reviewed. The higher molecular weight nylon 6 described in the work above was utilized in the study. The polymer nanocomposites were prepared with the DSM microcompounder that was described earlier. The control organomontmorillonites were prepared by:

- exchanging trimethyl hydrogenated tallow quaternary ammonium ion onto the montmorillonite;
- the utilization of Cloisite 20A;
- montmorillonite exchanged with the quats for the two previous controls but washed with methanol after the exchange.

The montmorillonite with lower exchange capacities was prepared by exchanging Li ions onto the montmorillonite at three different concentrations. The lithium-exchanged montmorillonites were heated at 80–300 °C in order to encourage the lithium to enter the octahedral structure of the montmorillonite and lower the effective exchange capacity.

The reduction of exchange capacities was reported to be 5%, 15%, and 30% lower. These montmorillonites with lower exchange capacities were exchanged with trimethyl hydrogenated tallow quaternary ammonium ion and dimethyl bis(hydrogenated tallow) quaternary ammonium ions (this is the quat treatment for Cloisite 20A) and subsequently washed with methanol after exchange. The two controls prepared with trimethyl hydrogenated tallow quat had different amounts of quat exchanged onto the montmorillonite. The organomontmorillonite prepared with the standard commercial method had approximately a 7 meq/100 g of montmorillonite higher quat content than the exchange capacity of the montmorillonite. The control washed with methanol had approximately a 10 meq/100 g of montmorillonite lower quat content than the exchange capacity of the montmorillonite. The quat content of the Cloisite 20A and the control washed with methanol were approximately the same and close to the exchange capacity of the montmorillonite.

The organomontmorillonites prepared from trimethyl hydrogenated tallow quat with the montmorillonites that had lower exchange capacities resulted in quat contents of approximately 62 meq/100 g montmorillonite for the lowest exchange capacity and unexpectedly similar quat contents for 5% and 15% reduced exchange capacity montmorillonites (approximately 76 and 77.5 meq/100 g of montmorillonite respectively).

The reduced exchange capacity montmorillonites exchanged with dimethyl bis(hydrogenated tallow) quats had the expected quat content as a function of reduced exchange capacity (approximately 80, 75, and 65 meq/100 g of montmorillonite for the 5, 15, and 30% reduced exchange capacity).

Evaluation of exfoliation efficiency was done by TEM and WAXS, using the same protocol as described earlier. The Young's modulus of the polymer composites was performed also using the same protocol as described earlier. The results with the organomontmorillonites prepared with the dimethyl bis(hydrogenated tallow) quat were, as expected from the work above, all inferior.

The evaluations with the organomontmorillonites exchanged with trimethyl hydrogenated tallow quat provided some unexpected results. The two controls provided the best performance. Interestingly, the control that was washed with methanol and had the lowest quat content was superior to the control with the higher quat content. The next best performer was the organomontmorillonite with the least quat content and the lowest exchange capacity. The poorest performers were the organomontmorillonites with 5% and 15% lower exchange capacities. They had similar quat contents and similar performances. It is difficult at this point in the analysis of the above information to identify the significant independent variables responsible for this unusual behavior.

6.4 Examination of the historical revelation of polymer–clay nanotechnology

Cho and Paul [13] provide a survey of the performance of the melt-blended nylon 6–organomontmorillonite nanocomposite prepared in their laboratory in relation to the nylon 6–montmorillonite nanocomposites prepared by *in situ* polymerization [14–17]. The *in situ* polymerization work provided by Toyota involves the exchange of 12-aminolauric acid onto the surface of the montmorillonite with subsequent polymerization with ε-caprolactam to produce the nylon 6–montmorillonite nanocomposite. The end groups of this nylon 6 nanocomposite will have a higher carboxylic acid content as a function of the organomontmorillonite content because of the initiation of polymerization by the carboxyl groups on the montmorillonite surface. This is similar to the increased carboxylic acid content from the polymerization synthesis of the amorphous nylon evaluated above. The twin screw extruder that was utilized for this work was the Haake described above. The nylon 6 was the higher molecular weight version.

The organomontmorillonite that was evaluated was prepared from the exchange of methyl bis(hydroxyethyl) rapeseed quaternary ammonium ions onto the surface of the montmorillonite. The exfoliation efficiency of the organomontmorillonite in the nylon 6 was determined by WAXS and TEM as described above. The Young's modulus was measured using the same protocol as described in the previous work described above. The modulus values were expressed as the ratio of the modulus of the polymer nanocomposites to the modulus of the pure polymer. This modulus ratio was plotted as a function of montmorillonite content in the polymer.

The exfoliation of the montmorillonite into the nylon 6 by melt blending was excellent. The modulus values of the polymer–montmorillonites prepared in this study compare favorably with the Toyota work. Hence, when montmorillonite is properly exfoliated and aligned in nylon 6 by melt blending or *in situ* polymerization, the Young's modulus values are comparable and predictable based on the modeling studies described in chapter 5.

In addition to the *in situ* polymerization method described above, anionic ring opening polymerization of ε-caprolactam can be utilized to prepare montmorillonite–nylon 6 polymer nanocomposites by *in situ* processing [18,19]. The polymerization process involves exfoliating an organomontmorillonite into molten ε-caprolactam and catalyzing the ring-opening polymerization with the sodium salt of ε-caprolactam. The work by Liu *et al.* [18] employs 2,4-diisocyanate as a co-catalyst; the work by Ming *et al.* [19] utilizes N-acetyl caprolactam as the co-catalyst. Montmorillonite for both investigations was obtained from the same supplier (Zhejiang Finghong Clay Chemicals Co. Ltd). The exchange capacity was lower (85 meq/100 g of montmorillonite) than the montmorillonite utilized from Southern Clay Products (Cloisite).

The work by Liu *et al.* exchanged the montmorillonite with dioctadecyl dimethyl quaternary ammonium ions; the work by Ming *et al.* utilized octadecyl trimethyl quaternary ammonium ions to exchange on the montmorillonite. Both organomontmorillonites seem to exfoliate well in the molten ε-caprolactam to produce exfoliated polymer nanocomposites.

A more practical approach to the production of polymer nanocomposites by anionic ring-opening polymerization was provided by Ming *et al.* The polymerization occurred in a twin-screw extruder that provided the shear for mixing and the temperature for polymerization. The molecular weight of the polymer in the nanocomposite was a function of residence time (screw speed) in the extruder. The M_n rose to a maximum of about 27 000 with increasing screw speed and then decreased to about $M_n = 20\,000$. The development of molecular weight seems to be independent of organomontmorillonite content up to 3 wt.%. The modulus was measured on fibers that were spun and drawn (1:3.5 draw ratio) from the polymer nanocomposite produced in the extruder. The modulus increases for 1% and 3% loading of montmorillonite in the nylon 6 were a modest 18.4% and 29.3% increase when compared to the pure nylon 6. Further work is necessary to determine why the compromise in Young's modulus is observed in these spun and drawn fibers when compared to the results of injection-molded nylon 6 nanocomposites.

Work by Shishan *et al.* [20] illustrates the importance of utilizing montmorillonite that has sodium as the counterion for the preparation of the organomontmorillonite for successful exfoliation by melt blending with a polymer. The montmorillonite in this work had calcium as the counterion. Calcium compromises the full exfoliation of montmorillonite in water. Hence, the organomontmorillonite prepared from octadecylamine hydrochloride does not fully exchange with montmorillonite.

The organomontmorillonite was washed repeatedly with hot aqueous ethanol solution which will extract additional octadecyl ammonium ion from the montmorillonite. The WAXS and TEM pictures clearly indicate compromised montmorillonite exfoliation in two different molecular weights ($M_n = 18\,000$ and 21 000) nylon 6 supplied by DSM.

The Molau test [40] for polymer–organomontmorillonite compatibility was performed by dispersing the nylon–organomontmorillonite composites in formic acid. The nylon and the organomontmorillonite demonstrated excellent compatibility with one another. This is consistent with the work above, which successfully produced exfoliated organomontmorillonite in nylon 6.

At this point in the discussion of the mechanical properties of montmorillonite–nylon 6 polymer nanocomposites, a review of the work carried out by the Toyota Central Research & Development Labs, Inc. that initiated an explosion of research and development of polymer–montmorillonite nanocomposite technology in relation to the subsequent work found above is provided. A series of three articles appeared in the *Journal of Materials Research* in 1993 in issue number 5.

The first article [21] discussed the preparation of organomontmorillonite with a series of ω-amino carboxylic acids that varied in chain length from 2 to 18 carbons. The montmorillonite was obtained from Kunimine Ind. Co. and later named as Kunipia F. The exchange capacity of the montmorillonite was measured at 119 meq/100 g of montmorillonite. The amino acids were reacted with HCl to produce the ammonium ions. The ammonium ions were exchanged onto the montmorillonite to prepare the organomontmorillonites. A relationship between the degree of exfoliation of montmorillonite and the chain length of the amino acids was determined (a longer chain length resulted in greater separation between the montmorillonite plates). The organomontmorillonites were found to incorporate readily into molten ε-caprolactam. The exfoliation of the organomontmorillonites in molten ε-caprolactam was a function of the chain length of the amino acid exchanged onto the montmorillonite (longer chain lengths resulted in greater exfoliation). This publication identified 12-aminolauric acid as the prime candidate for exchange onto the montmorillonite for subsequent polymerization with ε-caprolactam to produce nylon 6–montmorillonite polymer nanocomposites.

The second article [14] describes the *in situ* preparation of the nylon–montmorillonite polymer nanocomposite from the 12-aminolauric acid exchanged montmorillonite prepared in the previous article. The montmorillonite content in the polymerization varied from 1.5% to 59.6%. The general procedure was to melt ε-caprolactam, the organomontmorillonite prepared with 12-aminolauric acid, and 6-aminocaproic acid together with stirring at 100 °C for 30 min. The 6-aminocaproic acid was utilized to increase the rate of reaction when the organomontmorillonite content was less than 8%. Polymerization occurred at 250 °C over a 6 h period. The M_n decreased from 5 100 for a 1.5% montmorillonite loading to 3 200 for a 6.8% montmorillonite content. The montmorillonite appeared to be completely exfoliated, based on WAXS and TEM results. The carboxyl functional end groups on the nylon and nylon bound to the montmorillonite steadily increased with increased organomontmorillonite content. The carboxyl groups from the exchanged 12-aminolauric acid on the montmorillonite seem to serve as initiation sites for polymerization.

The last paper in this series [15] focused on the measurement of the mechanical properties of the nylon 6–montmorillonite nanocomposites prepared above. The control nylon 6 (1013B, Ube Industries) was reported to have a $M_w = 13\,000$. The montmorillonite content in the nylon 6 polymer nanocomposites varied from 1.9% to 7.1%. In the previous article, the montmorillonite content at 1.5% in the nylon 6 nanocomposite produced $M_w = 62\,000$; the nylon 6 nanocomposite at 6.8% produced $M_w = 29\,000$. The influence of molecular weight on the Young's modulus was not compensated for in the comparisons of the pure nylon with the nylon nanocomposites. The Young's modulus values were measured at 23 °C and 120 °C. The modulus values at 120 °C increased from about 0.19 GPa for pure nylon to about 0.7 GPa for the nylon 6 nanocomposite with 6.8% montmorillonite content. The heat distortion temperature climbed from approximately 65 °C to approximately 150 °C for the polymer nanocomposite with a 6.8% montmorillonite content. The authors argue the applicability of the "mixing law" (Equation 5.3) coupled with a constrained polymer region associated with the montmorillonite as the mechanism for reinforcement. Identification of the proper mechanism for reinforcement of nylon 6–montmorillonite is provided above by Paul *et al.* [5].

A second series of two articles was published by Toyota Research in the same year, 1993, in the *Journal of Polymer Science: Part A: Polymer Chemistry*. The first article [23] was similar in theme to the above work published in *Journal of Materials Research*. The montmorillonite was intercalated with ε-caprolactam in an aqueous HCl solution. This organomontmorillonite was treated in the same way as the 12-aminolauric acid-exchanged montmorillonite in the above work: by blending it with ε-caprolactam and aminocaproic acid and producing the polymer nanocomposite at 160 °C for 6 h. The montmorillonite content varied from 1.9% to 7.1%. The Young's modulus and heat distortion temperature of the polymer nanocomposite prepared with ε-caprolactam intercalated montmorillonite were reported to be improved, as compared with the polymer nanocomposite prepared with 12-aminolauric exchanged montmorillonite. Examination of the data indicates that the polymer nanocomposite with the ε-caprolactam intercalated montmorillonite had a higher montmorillonite content (5.3% versus 4.7%) and a higher molecular weight ($M_w = 19\,700$ versus 16 300) in relation to the polymer nanocomposite prepared with the 12-aminolauric acid exchanged montmorillonite. This discrepancy could easily explain the differences in performance.

The second article prepared the nylon 6–montmorillonite nanocomposite by adding ε-caprolactam, phosphoric acid, and 6-aminocaproic acid to a 3.2% aqueous dispersion of montmorillonite [24]. This mixture was heated to 260 °C to produce the polymer nanocomposite! Phosphoric acid is key to the successful exfoliation of the montmorillonite into the polymer matrix during this synthesis. Other inorganic and organic acids that were utilized in the synthesis resulted in various degrees of exfoliation and intercalation of the montmorillonite.

Phosphate is special in its relationship with montmorillonite. Phosphate attaches tenaciously to the edges of montmorillonite. This behavior is apparently significant to the development of fully exfoliated montmorillonite by this synthetic process. A comparison of the Young's modulus measured at 120 °C of the nylon 6–montmorillonite nanocomposite with the *in situ* nanocomposite prepared with montmorillonite exchanged with 12-aminolauric acid indicates that this synthesis has a slightly higher value. This synthesis produced a higher molecular weight; this could explain the increase.

A sixth article on modulus performance of nylon 6–clay nanocomposites was also published in 1995 in the *Journal of Applied Polymer Science* [25] by Toyota Research. The Young's modulus of nylon 6–clay nanocomposites was prepared with montmorillonite, saponite, a synthetic hectorite (Laponite RD), and synthetic mica. The *in situ* polymerization procedure with 12-aminolauric acid exchanged onto the minerals was utilized to prepare the nylon 6 nanocomposites. The montmorillonite polymer nanocomposite has the largest modulus, followed by mica, saponite, and, the lowest value, Laponite RD.

This work is not definitive because the aspect ratios and particle morphologies of all these minerals differ (for example, the aspect ratio of Laponite RD is about an order of magnitude smaller than the other minerals). WAXS was not provided for the clay nanocomposites;

only TEMs were published. Various extents of exfoliation as a function of mineral choice are suspected.

This early work by Toyota Research produced a spark that led to an explosion of exploration into the nature of nylon 6–montmorillonites that is summarized above. In addition, every commercial polymer in the world today has been evaluated in the context of montmorillonite nanocomposite preparation. A summary of the mechanical properties of those polymers is found below.

6.5 Examination of polymer–clay composites with complex processing issues

Work by Fornes and Paul [22] indicates that melt blending of organo-montmorillonite in an extruder can successfully exfoliate and align montmorillonite particles in a thermoplastic polymer (nylon 6) with the resulting dramatic improvement in mechanical properties in the direction of the montmorillonite particle alignment. However, not all polymers are amenable to this process. Epoxies are high-viscosity polymer liquids before cure that present challenges to exfoliation and subsequent alignment of the montmorillonite particles in the final cured state.

Koerner *et al.* [26] were successful in aligning exfoliated montmorillonite particles in epoxy with the utilization of a magnetic field. Two sources of naturally occurring montmorillonite were evaluated. The montmorillonite source from Nanocor was mined in South Dakota; the montmorillonite from Southern Clay Products was mined in Wyoming. Even though the elemental analysis of the two montmorillonites was similar, their magnetic properties were significantly different. The montmorillonite from Nanocor aligned perpendicular to the magnetic field and the montmorillonite from Southern Clay Products aligned parallel with the magnetic field.

Because the exfoliation of the montmorillonite in the epoxy was complete, SAXS and TEM were employed to characterize the orientation of the montmorillonite in the epoxy. When Epon 862 was evaluated with Epikure W as the curing agent, the exfoliation of organomontmorillonite (montmorillonite exchanged with octadecylammonium chloride at the exchange capacity) was excellent. The coefficient of thermal expansion (CTE) of the epoxy–montmorillonite nanocomposites was evaluated at 150 °C (a temperature above the T_g of the epoxy). When montmorillonite was aligned in the polymer by the magnetic field, the CTE decreased by an additional 20–30% in the direction of the particle alignment. Because the alignment of the montmorillonite from Nanocor and Southern Clay Products in the magnetic field was orthogonal, a mixture of the two montmorillonites provided CTE improvements in orthogonal orientations of the epoxy nanocomposite.

An example of the difficulty in dispersing organomontmorillonite into epoxy is illustrated in the work of Liu *et al.* [27]. They employed a high shear environment (a homogenizer at 10 000 r/min for 60 min and sonication for 60 min). The epoxy was DER 354 from Dow; the organoclay was Cloisite 93A from Southern Clay Products. Cloisite 93A is prepared by exchanging dihydrogenated tallow, methyl amine that is neutralized with HCl to prepare the ammonium ion onto montmorillonite near the exchange capacity (90 meq/100 g). The TEM pictures indicate a nonaligned composite with the poorly exfoliated montmorillonite. The low value for the storage modulus as a function of montmorillonite loading (10% increase at 4 wt.% loading of organomontmorillonite) supports the TEM observations. Fracture toughness of the epoxy increased with montmorillonite concentration, as expected with any dispersed-phase addition to epoxy. An additional concern in utilizing protonated amines for exchanging onto montmorillonite is discussed in Park and Jana [28], Brnardic *et al.* [29], and Salahuddin [30].

Epoxy ring opening is catalyzed by acids and bases. Evaluations of the role of ammonium ions as weak acids indicate that they do catalyze the epoxy cure. This can lead to a lowering of the T_g of the polymer. This compromises mechanical performance. TEM and mechanical properties indicate that the exfoliation and particle alignment of montmorillonite in the epoxy polymer in the above references are also significant variables.

Increases in shear intensity for the dispersion of montmorillonite into epoxy offer incremental improvements in exfoliated and aligned particle composites. A Lightnin LabMaster SI Mixer was evaluated for the dispersion of Cloisite 30B into EPON 828 [31] and a more effective method, ball milling, was evaluated with montmorillonite exchanged with dodecyl benzyl dimethyl quaternary ammonium ions and *meta*-xylenediamine [32,33].

Ball milling is time-consuming (10 h for the references cited is representative) but has a good reputation for preparing excellent pigment dispersions for inks and coatings. The ball-milling examples were further enhanced by lowering the viscosity of the epoxy dispersion through the dilution of the montmorillonite–epoxy dispersion with acetone. The composites prepared by ball milling demonstrated the largest improvement in mechanical properties (50% increase in impact strength) at a 3% loading of montmorillonite in the epoxy.

Full exfoliation and particle alignment continued to be problematical with these higher shear approaches. Work by Pluart *et al.* [34] provides additional support for the work above. This work eliminated high shear for the preparation of the montmorillonite–epoxy dispersion as a significant independent variable. The dispersions were prepared by hand-mixing. The organomontmorillonites in this study were Tixogel MP 250 (montmorillonite that is exchanged with the benzyl quat mentioned above) and montmorillonite exchanged with octadecylammonium ion (a similar treatment to that on Cloisite 93A cited above, except without the two methyl groups on the nitrogen). The epoxy in the study was LY 556 supplied by Vantico. The curing agents were Jeffamine D 2000 and 4,4′-methylenebis[3-chloro-2,6-diethylaniline] (DGEBA). Each curing agent was evaluated separately for each formulation. The loading levels are reported at parts per hundred parts of resin (polymer), phr, and not on a wt.% basis.

The intercalation of the epoxy into the galleries of the benzyl quat was consistent with the work above. The gallery spacing of the organomontmorillonite prepared with the primary amine ammonium ion expanded in the epoxy passed the capability of WAXS measurements. SAXS indicated an exfoliation of 7.0–11.0 nm at the end of the polymerization as a function of curing agent (D 2000 provided the smaller gallery spacing). The TEM pictures further support the problematical issues associated with aligning and dispersing exfoliated montmorillonite in epoxy.

The Young's modulus values for the Tixogel–epoxy composite at 5 phr cured with DGBA were about 60% greater than the amine-exchanged montmorillonite and fumed silica (as the control) at the same loading levels with the same curing agent. The apparent compromising effect of the catalysis of the epoxy ring opening by the protonated amine on mechanical properties was demonstrated again.

Epoxies have been toughened with elastomer-dispersed phases with a decrease in modulus. Work by Balakrishnan *et al.* [35] utilized organomontmorillonite as a dispersed phase in elastomer-toughened epoxies to recover this lost modulus. The montmorillonite was modified with octadecyl ammonium ion. The elastomer-dispersed phase in the epoxy was prepared by free radical polymerization of acrylic monomer within the epoxy. The acrylic elastomer-dispersed phase had epoxy functionality provided through the utilization of glycidyl methacrylate as a comonomer.

This "preformed" elastomer-reinforced epoxy is blended with additional epoxy and the organomontmorillonite to produce the final composite. The epoxy was DER 331 from Dow; the curing agent was piperdine. The formulations were based on phr. The authors did convert the montmorillonite content to wt.% when illustrating the results.

The difficulty in preparing aligned and exfoliated montmorillonite composites in epoxy was again apparent. This complication was further compromised with the preferred association at the epoxy–elastomer interface of the intercalated organomontmorillonite bundles. The result was a greater randomization of the montmorillonite dispersed phase in the epoxy.

The composite that contained 16 phr of elastomer and 5.5 wt.% of organomontmorillonite provided unexpected results. The Young's modulus recovered modestly when the montmorillonite was in the formula. The unexpected result was the increase of percent elongation to failure from 10.0% for the elastomer toughened epoxy to 11.5% when the montmorillonite was present.

Stress whitening (indicating cavitation during stress) for the montmorillonite containing elastomer-toughened epoxy was much more extensive that the stress whitening with the elastomer-toughened epoxy without montmorillonite. This unusual behavior will be discussed in Chapter 7 with the review of montmorillonite–elastomer nanocomposites and an unusual amorphous polyester nanocomposite, terephthalate copolyester prepared with 2,2,4,4-tetramethyl-1,3-cyclobutanediol and propanediol.

6.6 Polymer chain engineering in relation to montmorillonite incorporation as a nanoparticle

Another approach to the orientation of exfoliated montmorillonite in polymers is the utilization of block copolymers. Work by Yung-Hoon Ha *et al.* [36] prepared oriented, montmorillonite nanocomposites in a styrene-butadiene-styrene (SBS) block copolymer (Vector 4461-D). Attempts to prepare exfoliated montmorillonite in block copolymers have been unsuccessful until the advent of this work. The montmorillonite was surface treated by two methods.

One approach was the exchange of amine functional polystyrene onto the surface of the montmorillonite. The other approach was to surface treat the montmorillonite by growing polystyrene chains off the surface of the montmorillonite with the use of an initiator (2,2,6,6-tetramethyl-1-piperidinyloxy; TEMPO) that was associated with the surface of the montmorillonite. Two different molecular weights of polystyrene were evaluated for each surface modification strategy. The number-average molecular weights for the exchanged polystyrene were 1400 and 7100 g/mol; the number-average molecular weights of polystyrene that were polymerized onto the surface of the montmorillonite were 900 and 6000 g/mol. The block copolymer and block copolymer-nanocomposites were prepared by dissolving them into a tetrahydrofuran (THF) cumene solvent mixture and then evaporating the solutions to a viscous consistency. The viscous polymer solutions were roll cast and allowed to evaporate to their final state.

The SBS polymer has layers of polybutadiene sandwiched between layers of polystyrene. The orientation of the layers is orthogonal to the direction of roll casting. When a polymer nanocomposite was prepared with SBS that contained 2% montmorillonite that was surface modified with polystyrene at 6000 mol. wt. prepared by surface initiated polymerization, an excellent exfoliated montmorillonite product was prepared. The orientation of the SBS layers was now parallel to the direction of roll casting.

SAXS and TEM indicated that the montmorillonite was dispersed as single particles exclusively in the polystyrene layer. The thickness of the polystyrene layer increased from 12 nm to 25 nm at the location of the montmorillonite particle. The thickness of the polybutadiene layer was approximately 14 nm throughout the SBS polymer nanocomposite.

The SBS polymer nanocomposite layers were not as regular as those of the SBS polymer. Hence, the modulus of the SBS polymer nanocomposite was lower (120 MPa) than the SBS polymer (180 MPa). The yield stress was about 20% lower for the SBS polymer nanocomposite when compared to the SBS polymer. However, after the yield point, strain hardening occurred for the SBS polymer nanocomposite at a much lower percent elongation and the rate of strain hardening was higher when compared to SBS.

This approach to prepare fully exfoliated montmorillonite–polymer nanocomposites has yet to be fully exploited. One might imagine the unique mechanical performance of this type of block copolymer with different concentrations of montmorillonite in each layer.

Previous work by Ha and Thomas [37] provided the inspiration for the above work. The block copolymer for this work was polystyrene-isoprene-polystyrene (Vector 4411-D from Dexco Polymers) with a molecular weight of 18 000 for the styrene blocks and 44 000 for the isoprene block. The organomontmorillonite that was employed in the study was Cloisite 10A. The quat that is utilized to prepare the organomontmorillonite is dimethyl benzyl hydrogenated tallow quaternary ammonium ion. The milliequivalent ratio (MER) of quat associated with the montmorillonite in Cloisite 10A is 125. The exchange capacity of the montmorillonite is about 95. Hence, there are some 30 milliequivalents of quat per 100 g of montmorillonite that are not exchanged onto the montmorillonite.

The procedure for preparing the polymer composite involves dispersing the Cloisite 10A in toluene before blending the toluene dispersion with a toluene solution of the block copolymer. There will be a significant amount of quat that is dissolved in the toluene that is not associated with the montmorillonite (at least 30 milliequivalents). The toluene solution blend of Cloisite 10A and block copolymer is solvent cast and roll cast to prepare "isotropic" and "anisotropic" polymer composites.

The WAXS and TEM indicate some intercalation of polymer within the galleries of the montmorillonite. The Young's modulus for the solvent cast, isotropic, composite increased from 50 MPa for the pure block copolymer to 105 MPa for a 10 wt.% loading of Cloisite 10A. The roll-cast, anisotropic, polymer composite produced a Young's modulus increase from 153 MPa for the pure block copolymer to 165 MPa for the 10 wt.% Cloisite 10A loaded composite. The Young's modulus in the orthogonal direction to the polymer laminates was, as expected, lower. The pure block copolymer had a Young's modulus of 29 MPa in the orthogonal direction; the 10% loaded composite had a Young's modulus of 85 MPa in the orthogonal direction. SAXS evaluations indicated a synergistic relationship between the block copolymer orientations and the orientation of the Cloisite 10A dispersed phase.

In an attempt to improve the exfoliation efficiency of organomontmorillonite in block copolymers, Choi *et al.* [38] functionalized styrene-isoprene-styrene (SIS) and isoprene-styrene-butadiene (ISB) block copolymers with hydroxyl groups. The hydroxyl groups were formed by hydroboration with subsequent oxidation (OH^-, H_2O_2, H_2O). The hydroxyl groups will be on the primary carbon for terminal carbon–carbon double bonds. Hence, the ISB block copolymer will have terminal hydroxyl groups and the SIS block copolymer will have hydroxyl groups in the center of the block. For ISB, the molecular weight, Mn, was 9.6×10^3 with 34% styrene, 36% isoprene, and 30% butadiene. The SIS has a molecular weight, Mn, equal to 3.7×10^4 with a 46% content of styrene and 54% content of isoprene. The hydroxyl content of ISB was evaluated at 16, 24, and 39 mol %. The hydroxyl content of SIS was evaluated at

16 and 32 mol %. The two organomontmorillonites that were evaluated were Cloisite 30B and Cloisite 15A. As a reminder, Cloisite 30B has two primary hydroxyl groups associated with the quat; Cloisite 15A has an additional 30% excess dehydrogenated tallow dimethyl quat past the exchange capacity of the montmorillonite.

The block copolymers were dissolved and the organomontmorillonites were dispersed separately in the same cosolvent, a blend of tetrahydrofuran and water (a 90:10 mixture by volume). The reason for the unusual choice of cosolvent was not provided. Each polymer solution was blended with each organomontmorillonite dispersion. The solvents were allowed to evaporate over a period of two days.

The polymer composites, based on TEM and WAXS, indicated intercalated montmorillonite composites. The mechanical properties were determined with a parallel plate rheometer as a function of frequency and temperature. All the composites were evaluated at a 5% organomontmorillonite loading. As expected, the block copolymer with the highest concentration of terminal hydroxyl groups, ISB at 39 mol % hydroxyl content, produced the best dispersion of the more polar organomontmorillonite, Cloisite 30B. A comparison of the storage modulus as a function of temperature (about 70 °C to about 130 °C) for Cloisite 15A and Cloisite 30B at 0.1 rad/s produced almost identical results. Because the organic content of Cloisite 15A is higher than Cloisite 30B, the montmorillonite content was significantly less in the Cloisite 15A composite. Apparently, increasing the polarity of ISB and SIS block copolymers with hydroxyl functionality is not a panacea for producing fully exfoliated montmorillonite block copolymer nanocomposites.

Yamaguchi and Yamada [39] utilized the addition of a long chained fatty acid (stearic acid) to an organomontmorillonite (exchanged with octadecyl ammonium chloride) to increase the hydrophobic nature of the organomontmorillonite interface in an evaluation of exfoliation efficiency in a SBS block copolymer. The molecular weight of the block copolymer was Mn equal to 120 000. The styrene content was 30 wt. %.

The organomontmorillonite was prepared by dispersing Kunipia F (produced by Kunimine Ind. Co., Japan) in distilled water at 80 °C. An aqueous solution of HCl neutralized octadecyl amine and different concentrations of stearic acid were added to the heated montmorillonite aqueous dispersion. The amount of ammonium ion utilized to exchange onto the montmorillonite (CEC = 115 meq/100 g of montmorillonite) was close to the exchange capacity (120 meq). The amount of carboxylic acid that was added was 20, 50, and 100 meq. Low shear mixing

(Labo-Plastmil manufactured by Toyo Seiki Co., 130 °C, 30 r/min, for 3 min) was employed to prepare the polymer composite.

Compare the above reluctance to melt blending (all the block copolymer composite work was performed by solvent casting) to this work. The test samples from the melt-blended composites were prepared by pressing at 130 °C for 3 min with 10 MPa pressure. The shear from hot pressing will align the montmorillonite in the shear direction. The loading of the organomontmorillonites in the block copolymer was 5 and 10 wt.%. Note that the montmorillonite content in the copolymer will decrease as the organic content (increase in carboxylic acid content) of the organomontmorillonite increases. Intercalation of the organomontmorillonite with polymer occurs when the carboxylic acid is present; the organomontmorillonite without the carboxylic acid is poorly intercalated and poorly dispersed in the polymer.

The tensile modulus for the composites at 100% and 300% elongation reached a maximum with a carboxylic acid content of 20 meq in the organomontmorillonite. At 10% loading of organomontmorillonite with a 20 meq content of carboxylic acid, the tensile modulus increased from 1.69 MPa for pure copolymer to about 3.5 MPa at 100% elongation and increased from 2.04 MPa for the pure copolymer to about 4.63 MPa at 300% elongation.

Intercalation, improved montmorillonite dispersion, and alignment of the montmorillonite in the direction of stress during testing were all positive factors in the significant improvement of mechanical properties. Greater hydrophobicity at the montmorillonite surface through the utilization of a long hydrocarbon chained carboxylic acid with the quat did not overcome the barrier to full exfoliation in the block copolymer.

6.7 Conclusions

The introduction of montmorillonite as a nanoparticle into a polymer provides a grievous insult to the tranquility of the polymer chain–polymer chain interactions. A complete description of the significant independent variables associated with the process is found in Chapter 2. The most dire aspect of the process is the diffusion of the polymer chain into the narrow galleries between the montmorillonite plates.

The transition of the polymer chain from a three-dimensional environment found in the continuous phase to an effectively two-dimensional environment associated with the gallery can be mediated by proper engineering of the polymer backbone, utilization of the proper processing equipment and protocol for melt processing, and the adoption of solvent for polymer nanocomposite casting.

The adoption of altered polymer chain–polymer chain morphologies for polymer–montmorillonite nanocomposites, particularly for crystalline polymers, indicates the significance of the energetic compromises associated with nanocomposite construction. The success of the preparation of nylon 6 nanocomposites in relation to the difficulties associated with the preparation of nylon 6,6 nanocomposites illustrates the sensitivity of the preparation of polymer nanocomposites to the fundamental polymer structure–property relationships.

The twin-screw extruder is a necessity for the proper preparation of melt-blended polymer nanocomposites. They are not a panacea. The processing temperature, shear conditions, and residence time in the extruder are all significant independent variables associated with the successful preparation of polymer nanocomposites. The choice of the injector molder, conditions, and protocol are critical for the proper alignment of the montmorillonite particles in the direction of applied stress associated with part design.

Optimizing the hydrophilic–hydrophobic balance of the surface treatment of montmorillonite in the context of the hydrophilic–hydrophobic balance of the polymer continuous phase is also critical to the successful preparation of polymer–montmorillonite nanocomposites. Neither is this a panacea. Useful parameters (for example, solubility parameters) can be employed as guides for the choice of the proper surface modification of the montmorillonite.

The viscosity of the polymer continuous phase will compromise the successful exfoliation and alignment of montmorillonite in the composite. Clever utilization of techniques that take advantage of the unique character of montmorillonite (for example, magnetic properties) will resolve the difficulties associated with high viscosities. When all else fails, casting of the polymer and montmorillonite dispersions from solution results in the successful preparation of polymer–montmorillonite nanocomposites.

References

[1] F. Chavarria and D. R. Paul. Comparison of nanocomposites based on nylon 6 and nylon 66. *Polymer*, **45** (2004), 8501–8515.

[2] Y. Kojima, A. Usuki, M. Kawasumi, A. Okada, T. Kurauchi, O. Kamigaito, and K. Kaji. Fine structure of nylon-6–clay hybrid. *Journal of Polymer Science: Part B: Polymer Physics*, **32** (1994), 625–630.

[3] Y. Yoo and D. R. Paul. Effect of organoclay structure on morphology and properties of nanocomposites based on an amorphous polyamide. *Polymer*, **49** (2008), 3795–3804.

[4] T. D. Fornes, P. J. Yoon, H. Keskkula, and D. R. Paul. Nylon 6 nanocomposites: the effect of matrix molecular weight. *Polymer*, **42** (2001), 9929–9940.

[5] H. R. Dennis, D. L. Hunter, D. Chang, S. Kim, J. L. White, J. W. Cho, and D. R. Paul. Effect of melt processing conditions on the extent of exfoliation in organoclay-based nanocomposites. *Polymer*, **42** (2001), 9513–9522.

[6] J. M. Rubi. The long arm of the second law. *Scientific American*, Nov. (2008), 62–67.

[7] F. Chavarria, R. K. Shah, D. L. Hunter, and D. R. Paul. Effect of melt processing conditions on the morphology and properties of nylon 6 nanocomposites. *Polymer Engineering and Science*, **47**:11 (2007), 1847–1864.

[8] T. D. Fornes, P. J. Yoon, D. L. Hunter, H. Keskkula, and D. R. Paul. Effect of organoclay structure on nylon 6 nanocomposite morphology and properties. *Polymer*, **43** (2002), 5915–5933.

[9] S. Xie, S. Zhang, B. Zhao, H. Qin, F. Wang, and M. Yang. Tensile fracture morphologies of nylon-6/montmorillonite nanocomposites. *Polymer International*, **54** (2005), 1673–1680.

[10] D. W. Schaefer and R. S. Justice. How nano are nanocomposites? *Macromolecules*, **40**:24, (2007), 8501–8517.

[11] D. W. Schaefer, R. S. Justice, H. Koerner, R. Vaia, and C. Zhao. Large-scale morphology of dispersed layered silicates. *Materials Research Society Symposium Proceeding*, **840** (2005), 57–62.

[12] F. Chavarria, K. Nairn, P. White, A. J. Hill, D. L. Hunter, and D. R. Paul. Morphology and properties of nanocomposites from organoclays with reduced cation exchange capacity. *Journal of Applied Polymer Science*, **105** (2007), 2910–2924.

[13] J. W. Cho and D. R. Paul. Nylon 6 nanocomposites by melt compounding. *Polymer*, **42** (2000), 1083–1094.

[14] A. Usuki, Y. Kojima, M. Kawasumi, A. Okada, Y. Fukushima, T. Kurauchi, and O. Kamigaito. Synthesis of nylon 6-clay hybrid. *Journal of Materials Research*, **8**:5 (1993), 1179–1184.

[15] Y. Kojima, A. Usuki, M. Kawasumi, A. Okada, Y. Fukushima, T. Kurauchi, and O. Kamigaito. Mechanical properties of nylon 6-clay hybrid. *Journal of Material Research*, **8**:5 (1993), 1185–1189.

[16] Y. Kojima, A. Usuki, M. Kawasumi, A. Okada, T. Kurauchi, and O. Kamigaito. Synthesis of nylon 6–clay hybrid by montmorillonite intercalated with ε-caprolactam. *Journal of Polymer Science: Part A: Polymer Chemistry* **31** (1993), 983–986.

[17] A. Usuki, A. Koiwai, Y. Kojima, M. Kawasumi, A. Okada, T. Kurauchi, and O. Kamigaito. Interaction of nylon 6–clay surface and mechanical properties of nylon 6–clay hybrid. *Journal of Applied Polymer Science*, **55** (1995), 119–123.

[18] A. Liu, T. Xie, and G. Yang. Synthesis of exfoliated monomer casting polyamide 6/Na^+-montmorillonite nanocomposites by anionic ring opening polymerization. *Macromolecular Chemistry and Physics*, **207** (2006), 701–707.

[19] Z. Ming, P. Xiangquing, and Y. Wang. Preparation and characterization of polyamide6/montmorillonite nanocomposites by reactive extrusion. *International Journal of Polymer Materials*, **55** (2006), 147–159.

[20] W. Shishan, J. Dingjun, O. Xiaodong, W. Fen, and S. Jian, The structure and properties of PA6/MMT naonocomposites prepared by melt compounding. *Polymer Engineering and Science*, **44**:11 (2004), 2070–2074.

[21] A. Usuki, M. Kawasumi, Y. Kojima, A. Okada, T. Kurauchi, and O. Kamigaito. Swelling behavior of montmorillonite cation exchanged for ω-amino acids by ε-caprolactam. *Journal of Materials Research*, **8**:5 (1993), 1174–1178.

[22] T. D. Fornes and D. R. Paul. Modeling properties of nylon 6/clay nanoparticles using composite theories. *Polymer*, **44** (2003), 4993–5013.

[23] Y. Kojima, A. Usuki, M. Kawasumi, A. Okada, T. Kurauchi, and O. Kamigaito. Sythesis of nylon 6–clay hybrid by montmorillonite intercalated with ε-caprolactam. *Journal of Polymer Science: Part A: Polymer Chemistry*, **31** (1993), 983–986.

[24] Y. Kojima, A. Usuki, M. Kawasumi, A. Okada, T. Kurauchi, and O. Kamigaito. One-pot synthesis of nylon 6–clay hybrid. *Journal of Polymer Science: Part A: Polymer Chemistry*, **31**:7 (1993), 1755–1758.

[25] A. Usuki, A. Koiwai, Y. Kojima, M. Kawasumi, A. Okada, T. Kurauchi, and O. Kamigaito. Interaction of nylon 6–clay surface and mechanical properties of nylon 6–clay hybrid. *Journal of Applied Polymer Science*, **55** (1995), 119–123.

[26] H. Koerner, E. Hampton, D. Dean, Z. Turgut, L. Drummy, P. Mirau, and R. Vaia. Generating triaxial reinforced epoxy/montmorillonite nanocomposites with uniaxial magnetic fields. *Chemistry of Materials*, **17** (2005), 1990–1996.

[27] T. Liu, W. C. Tjiu, Y. Tong, C. He, S. S. Goh, and T.-S. Chung. Morphology and fracture behavior of intercalated epoxy/clay nanocomposites. *Journal of Applied Polymer Science*, **94** (2004), 1236–1244.

[28] J. Park and S. C. Jana. Adverse effects of thermal dissociation of alkyl ammonium ions on nanoclay exfoliation in epoxy-clay systems. *Polymer*, **45** (2004), 7673–7679.

[29] I. Brnardic, M. Ivankovic, H. Ivankovic, and H. J. Mercer. Isothermal and nonisothermal cure kinetics of an epoxy/poly(oxypropylene)diamine/octadecylammonium modified montmorillonite system, *Journal of Applied Polymer Science*, **100** (2006), 1765–1771.

[30] N. A. Salahuddin. Layered silicate/epoxy nanocomposites: synthesis, characterization and properties. *Polymers for Advanced Technologies*, **15** (2004), 251–259.

[31] T.-K Oh, M. Hassan, C. Beatty, and H. El-Shall. The effect of shear forces on the microstructure and mechanical properties of epoxy–clay nanocomposites. *Journal of Applied Polymer Science*, **100** (2006), 3465–3473.

[32] H.-J Lu, G.-Z. Liang, X.-Y. Ma, B.-Y. Zhang, and X.-B. Chen. Epoxy/clay nanocomposites: further exfoliation of newly modified clay induced by shearing force of ball milling. *Polymer International*, **53** (2004), 1545–1553.

[33] H.-J. Lu, B.-Y. Zhang, and X.-B. Chen. Effect of external shearing force on exfoliation structure and properties of high-performance epoxy/clay nanocomposites. *Transaction of the Nonferrous Metals Society of China*, **15**:2 (2005), 211–216.

[34] L. L. Pluart, J. Duchet, and H. Sautereau. Epoxy/montmorillonite nanocomposites: influence of organophilic treatment on reactivity, morphology and fracture properties. *Polymer*, **46** (2005), 12267–12278.
[35] S. Balakrishnan, P. R. Start, D. Raghavan, and S. D. Hudson. The influence of clay and elastomer concentration on the morphology and fracture energy of preformed acrylic rubber dispersed clay filled epoxy nanocomposites. *Polymer*, **46** (2005), 11255–11262.
[36] Y.-H. Ha, Y. Kwon, T. Breiner, E. P. Chan, T. Tzianetopoulou, R. E. Cohen, M. C. Boyce, and E. L. Thomas. An orientationally ordered hierarchical exfoliated clay–block copolymer nanocomposite. *Macromolecules*, **38**:12 (2005), 5170–5179.
[37] Y.-H. Ha and E. L. Thomas. Deformation behavior of a roll-cast layered-silicate/lamellar triblock copolymer nanocomposite. *Macromolecules*, **35** (2002), 4419–4428.
[38] S. Choi, K. M. Lee, and C. D. Han. Effects of triblock copolymer architecture and the degree of functionalization on the organoclay dispersion and rheology of nanocomposites. *Macromolecules*, **37** (2004), 7649–7662.
[39] T. Yamaguchi and E. Yamada. Preparation and mechanical properties of clay/polystyrene-block–polybutadiene-block–polystyrene triblock copolymer (SBS) intercalated nanocomposites using organoclay containing stearic acid. *Polymer International*, **55** (2006), 662–667.
[40] G. E. Molau. Heterogenous polymer systems. VI. Influence of molecular weight on poo [polymeric oil-in-oil] emulsifier activity of SB [styrene-butadiene] block copolymers, *Kolliod Zeitschrift & Zeitschrift fuer Polymere*, **238** (1–2) (1970), 493–398.

7 The relationships of polymer type specificity to the production of polymer–clay nanocomposites

7.1 Complexity of polyolefin–montmorillonite nanocomposites

Preparing polyolefin–montmorillonite nanocomposites presents another challenge in relation to the preparation of block copolymer–montmorillonite nanocomposites found in Chapter 6. An excellent example of the complexity of exfoliating organomontmorillonite into a pure hydrocarbon polymer is found in the work by Hotta and Paul [1]. Linear low-density polyethylene (LLDPE; Dowlex 2032 manufactured by Dow Chemical) was melt blended with two different organomontmorillonites (Cloisite 20A and montmorillonite exchanged with trimethyl hydrogenated tallow quaternary ammonium ion). The importance of blending maleic anhydride grafted LLDPE (LLDPE–g–MA; 0.9 wt.% MA content; Fusabond MB266D produced by DuPont, Canada) with LLDPE as regards achieving exfoliation was determined in this study.

The procedures and equipment that were employed in this work were identical to those utilized by Fornes and Paul in the preparation of melt-blended nylon 6–montmorillonite nanocomposites described in Chapter 5. As one may anticipate from the studies in Chapters 5 and 6, the more hydrophobic Cloisite 20A was more efficient in producing exfoliated composites. The presence of the LLDPE–g–MA in the polymer blend further encouraged the exfoliation of Cloisite 20A. When the weight ratio of LLDPE–g–MA to Cloisite 20A is increased to 4 and subsequently to 11, the WAXS indicated good exfoliation with a loading of 4.6 and 4.9 wt.%, respectively, of montmorillonite (determined by incineration of the polymer composite in an oven). The TEM for the composite with a ratio of 11 at 4.9% montmorillonite indicated good exfoliation.

The tensile modulus of LLDPE with a loading of 4.2 to 4.9 wt.% of Cloisite 20A as a function of LLDPE–g–MA provided surprising results. As the weight ratio of LLDPE–g–MA to Cloisite 20A increases from 0 to 1, the tensile modulus increases from 413 to 480 MPa. As the ratio increases from 2 to 11, the tensile modulus decreases to 399 MPa. The above WAXS and TEM evaluations indicate an increase in exfoliation efficiency as a function of increased LLDPE–g–MA content for Cloisite 20A. The models developed in the first part of Chapter 5 predict a steady increase in tensile modulus as a function of increased exfoliation of Cloisite 20A. Blends of LLDPE with LLDPE–g–MA without Cloisite 20A content mimic this behavior at lower tensile modulus values (190 MPa for pure LLDPE climbing to a maximum of 216 MPa at low LLDPE–g–MA content and falling again to 153 MPa for pure LLDPE–g–MA). LLDPE–g–MA has a lower tensile modulus than LLDPE, presumably because of the decreased crystallinity of the polymer. Referring again to the predictive models developed in the first part of Chapter 5, increased exfoliation of Cloisite 20A in polymer blends of LLDPE and LLDPE–g–MA apparently compromises the tensile modulus of the continuous phase (the polymer blend).

A further complication of preparing LDPE–montmorillonite nanocomposites is the apparent effect of the thermal degradation of the quat on the montmorillonite during processing. Shah and Paul [2] utilized LD 621 (produced by Exxon Mobil Chemical) as the LDPE for melt blending with Cloisite 20A in a temperature range from 150 to 240°C in the same equipment that was employed above. The processing temperature for the above study was 200°C. The montmorillonite loading level in the polyethylene was 5%. WAXS indicated intercalated composites (fulfilling expectations from the above work). No TEM work was reported.

When the process temperature for the preparation of the polymer composites reached 200°C, WAXS indicated a decrease in gallery spacing between the montmorillonite plates. This gallery spacing decreased further at the processing temperature of 240°C. The ratio of the tensile modulus of the polymer composites divided by the modulus of the pure polymer begins decreasing at 180°C from 2.5 to 2.3 and continues decreasing with increased processing temperature to 1.9 at 140°C. Isothermal thermogravimetric analysis (TGA) indicates significant thermal degradation of Cloisite 20A and the polymer composites prepared with Cloisite 20A over this temperature range. The question remains: could thermal degradation of the polymer composite with Cloisite 20A in the above work be a significant variable in compromising the tensile modulus of the exfoliated composites?

Quantification of the role of polar groups associated with polyethylene in exfoliation of organomontmorillonite was carried out by Shah *et al.* [3]. The polar modification to polyethylene was methacrylic acid (MAA) copolymerized with the ethylene monomer (Surlyn 8920; 5.53 mole % MAA, 8940; 5.81 mole % MAA, and 8945; 5.59 mole % MAA; manufactured by DuPont). A similar survey of different quaternary ammonium ions exchanged on montmorillonite found in Chapter 5 with nylon 6 [4] was duplicated in this study. The exceptions were the evaluation of Cloisite 6A in this study as opposed to Cloisite 15A in the nylon 6 study and montmorillonite exchanged with methyl trihexadecyl quaternary ammonium ion. The montmorillonite in this study has an exchange capacity of approximately 95 meq/100g of montmorillonite. Cloisite 6A has 140 MER of dimethyl dehydrogenated tallow associated with it. Hence, there is approximately 45 meq/100g of montmorillonite of unexchanged quat associated with the organomontmorillonite.

This positively charged quat has the potential to form an organic salt with the negatively charged carboxyl groups associated with the polyethylene. The formation of organic salts is a common occurrence in organic chemistry. The formation of these salts will increase the hydrophobicity of Surlyn and change the morphology of the polyethylene with long hydrocarbon chains pendant to the main polyethylene chain.

An initial survey of the three Surlyn polymers with Cloisite 20A indicated that 8945 was the best performer. Subsequent survey work with the other organomontmorillonites focused exclusively on 8945. The same equipment and analysis protocol was employed for this survey as that utilized in the nylon 6 survey in Chapter 5. The loading of montmorillonite in the polymer varied from 2.5 wt.% to 10.0 wt.%. As expected from the above work, the organomontmorillonites exchanged with the most hydrophobic quats (Cloisite 20A, 6A, and methyl trihexadecyl quaternary ammonium ion) demonstrated superior exfoliation efficiency based on TEM and WAXS.

Work with Cloisite 30B demonstrated that the primary hydroxyls associated with bis(2-hydroxy-ethyl)methyl tallow quaternary ammonium ion exchanged onto montmorillonite was significantly beneficial for increasing exfoliation efficiency in 8945. The exfoliation efficiency of the best-performing organomontmorillonites in 8945 was inferior to the best exfoliation efficiency found in nylon 6. The increase in modulus from the pure 8945 (0.262 GPa) for the three best-performing nanocomposites at 10 wt.% loading were 0.919 GPa for Cloisite 20A, 1.008 GPa for Cloisite 6A, and 1.126 GPa for the methyl trihexadecyl quaternary ammonium ion exchanged montmorillonite.

Additional work seems to be suggested from this investigation to include a separate incorporation of different quats associated with the carboxyl groups of 8945 in relation to an independent incorporation of different organomontmorillonites and the integration of hydroxyl functionality into a hydrophobic quat (for example, dimethyl dehydrogenated tallow quaternary ammonium ion and methyl trihexadecyl quaternary ammonium ion) in order to test the potential synergistic effects of these significant independent variables.

The significance of the counterion identity associated with Surlyn with regard to exfoliation efficiency and mechanical properties was explored by Shah and Paul [5]. The counterions were lithium (Surlyn 9945), sodium (8945 evaluated above), and zinc (Surlyn 7940). Each copolymer was neutralized to 40%. The procedures for preparation of the polymer nanocomposites and characterization were identical to that described above, except that the feed rate for the extruder was decreased to 800 g/h because the high build in viscosity associated with utilization of the zinc counterion 9945. This prevented the power capabilities of the extruder from being exceeded. Cloisite 20A was utilized as the organomontmorillonite for the preparation of the polymer composite. The loading levels of organomontmorillonite in the polymers were 2.5, 5.0, and 10.0 wt. %.

The lithium counterion polymer demonstrated the poorest exfoliation efficiency. The sodium and zinc counterion polymers seem to demonstrate equally good exfoliation. The tensile modulus for each of the composites reflected this behavior (better exfoliation resulted in higher modulus).

It is worth reflecting on the different characteristics of these counterions at this point. Lithium has the highest charge density (smallest size) of the three counterions evaluated. It is extremely difficult to remove all of the hydration water from this ion. Lithium is expected to form an intimate association with carboxyl groups. Zinc ion is effective at forming coordination complexes with carboxylate ions.

Commercial coating successes utilizing this effect have been available for a number of years. This effect in coatings is referred to as ionic crosslinking. For example, Johnson wax products (manufactured by S. C. Johnson) are formulated with zinc ammonia complexes in water-based coatings that contain carboxylate functionality associated with acrylic polymers. As the coating dries, the ammonia is fugitive. The carboxylate functionality of the polymer replaces the ammonia, to form a coordination complex with zinc. The result is a tough and durable coating on a substrate (the floor).

Sodium has the lowest charge density of these three ions and demonstrates none of the characteristics specifically associated with the other two counterions. An examination of the ratio of the modulus of the composites divided by the modulus of the pure polymer indicates a 4.51 factor increase for the zinc counterion composite at 10% loading; the sodium counterion composite at 10% loading has a lower 3.49 factor increase. As expected (based on lower exfoliation efficiency), the lithium counterion polymer at 10% loading produced a 2.32 factor increase in modulus. It is anticipated that an optimization of zinc counterion content in the polymer, coupled with the carboxylate level, would give a better mechanical performance. It is tempting to suggest that ionic cross-linking by zinc with carboxyl functionality is synergistic with the reinforcement mechanism provided by montmorillonite in these polymer nanocomposites.

Shah *et al.* [6] evaluated polyethylene–MAA copolymers without neutralized MAA (Nucrel 0403, 3.9 wt. % MAA and Nucrel 0903, 8.9 wt. % MAA; manufactured by DuPont) in relation to Surlyn 8945 with 40% of the MAA neutralized with sodium counterions. The organomontmorillonite with the best exfoliation performance in this study was Cloisite 20A. The loading levels of Cloisite 20A were 2.5, 5.0, 7.5, and 10 wt.%. The processing and characterization protocol were identical to that employed above. Because the zinc counterion, Surlyn, with the high melt viscosity, was not evaluated in this study, the feedrate was increased back to the standard 1200 g/h that was utilized in the previous studies above. The total MAA content of 8945 is 15.2 wt.%. Hence, the amount of un-neutralized MAA in 8945 is approximately the same as 0903.

The TEM and WAXS analysis indicates that Cloisite 20A is more fully exfoliated in 8945. Cloisite 20A does exfoliate more efficiently in 0403 and 0903 when compared with the low-density polyethylene homopolymer, LD 671 manufactured by Exxon Mobil, utilized in this study. The aspect ratios of the montmorillonite at 2.5 wt.% loading in 0403 and 0903 determined by TEM evaluations were very similar (the number-average aspect ratio was 26 for 0403 and 28 for 0903).

It is temping to conclude that the MAA concentration difference in these two copolymers is not a significant independent variable as regards the exfoliation efficiency of Cloisite 20A in these copolymers with the processing conditions employed. The aspect ratio of Cloisite 20A at 2.5 wt.% loading in 8945 was 36. Is the higher MAA concentration in 8945 or the 40% neutralization of the MAA in 8945 responsible for the improved exfoliation efficiency?

The following work is focused on deconvoluting these issues. Because the tensile modulus differences between the polyethylene and the copolymers without montmorillonite are significant (LD 621 = 114 MPa, 0403 = 118 MPa, 0903 = 73 MPa, and 8945 = 262 MPa), one must be cautious about attributing changes in tensile modulus as a function of montmorillonite content to the degree of exfoliation. The theories developed in Chapter 5 predict that the polymers with the lowest modulus will demonstrate the greatest change in modulus as a function of montmorillonite content at the same degree of exfoliation (aspect ratio) because the difference in the modulus of the continuous phase (polymer) and the dispersed phase (montmorillonite) is greater. The tensile modulus of 8945 with 10 wt.% loading of Cloisite 20A has the largest value (919 MPa), but the ratio of the tensile modulus of 0903 at 10 wt.% Cloisite 20A divided by the tensile modulus of the pure copolymer has the largest relative modulus (4.83).

Work reported by Cui *et al.* [7] quantifies the role of MAA neutralization as regards the efficiency of exfoliating Cloisite 20A in ethylene-MAA copolymers. Nucrel 925 (manufactured by DuPont) is an ethylene-MAA copolymer that has an un-neutralized MAA content of 15 wt.%. This copolymer was compared with experimental copolymers prepared with 15 wt.% MAA with the degree of neutralization varying from 20, 30, 40, 50, 60, and 70%. The counterion was sodium. 925 is not the copolymer that was neutralized. The molecular weight of 925 was higher than the copolymer that was neutralized. The tensile modulus of the copolymers without added montmorillonite climbed from approximately 0.06 GPa for 925 with increasing neutralization to a plateau of about 0.36 GPa starting at approximately 60% neutralization. Cloisite 20A polymer nanocomposites were prepared and characterized utilizing an identical protocol as the work above. The extruder processing temperature was increased from 200 to 220°C as a function of increasing neutralization to compensate for the steady increase in the viscosity of the polymer melt. The loading levels of montmorillonite in the copolymers were 2.5, 5, and 7.5 wt.%.

TEM and WAXS results clearly indicate that the copolymer with the highest degree of neutralization produced the highest degree of exfoliation. The highest degree of exfoliation was inferior to that achieved with nylon 6. The tensile modulus for the copolymer that demonstrated the highest degree of exfoliation (70% neutralization) increased from approximately 0.36 GPa without the presence of montmorillonite to 0.8 GPa at 7.5% loading.

Additional work to identify the importance of polar modification of polyethylene in regard to organomontmorillonite exfoliation was carried out by Cui *et al.* [8] with polyethylene–vinyl acetate copolymers. Note that the acetate functionality on polymers will readily hydrolyze to a pendant hydroxyl group on the polymer and acetic acid at ambient conditions. Hence, the hydroxyl content of the copolymer can vary with time and conditions.

Low-density polyethylene was utilized as the control (LF-0219-A; Novapol). The ethylene–vinyl acetate copolymers varied in vinyl acetate content from 9.3, 18, 28, to 40% (Elvax 760Q, 460, 265, and 40L-03, respectively; DuPont). The organomontmorillonites that were evaluated were Cloisite 20A, Cloisite 30B, and montmorillonite exchanged with trimethyl hydrogenated tallow quaternary ammonium ion at the exchange capacity of the montmorillonite.

The procedure for the preparation of the polymer nanocomposites was identical to that found above. The protocol for characterizing the nanocomposites and the evaluation of the tensile modulus was identical to that employed for the above work. Cloisite 20A gave a superior performance in the polyethylene–vinyl acetate copolymers when compared to the other organomontmorillonites. The composites with 40% vinyl acetate content were not evaluated for mechanical properties because of the difficulty in preparing samples for testing. As the vinyl acetate content of the copolymer increased, the exfoliation efficiency of Cloisite 20A increased along with the mechanical performance. As the vinyl acetate content increased in the copolymer, the tensile modulus steadily decreased for the pure copolymer. This is understood to be the result of the steady decrease in crystallinity as a function of increasing vinyl acetate content in relation to the pure polyethylene. Because of this disparity of modulus as a function of composition, the ratio of the modulus of the composites as a function of organomontmorillonite content divided by the modulus of the pure copolymer was a useful metric for comparing composites.

The montmorillonite contents in the copolymers varied by 1, 3, 5, and 7 wt.%. The relative modulus of the Cloisite 20A composites with 18 and 28% vinyl acetate content increased to a factor of approximately 7 with 7% montmorillonite content. The percent elongation to failure for the composites prepared with the copolymer with 18% vinyl acetate also increased from 236% for the pure copolymer to approximately 270% for the composite with 3% montmorillonite.

This phenomenon of increased percent elongation of polymer composites as a function of increasing montmorillonite content will be addressed later in the chapter when the same behavior is observed with an unusual amorphous polyester and elastomers. This behavior is not observed with other dispersed-phase reinforcements. An evaluation of the results of the study with Cloisite 20A with the predictions of the Halpin–Tsai model that was introduced in the early part of Chapter 5 resulted in reasonable agreement. The exfoliation of Cloisite 20A in the copolymers with the highest vinyl acetate content was good but not as good as the exfoliation of the best-performing organomontmorillonite in nylon 6 determined in Chapter 5.

Suh *et al.* [9] explored the synergistic effects of blending EVA copolymers with polyethylene maleic anhydride (PEMA) to achieve exfoliation of montmorillonite. The organomontmorillonite in the study was Cloisite 15A. The vinyl acetate content of the EVA was evaluated at 3 (EF221), 8 (EF321), 12 (EF443), (manufactured by Hyundai Petrochemical Co.), 15 (E153F), 22 (E220F), (manufactured by Samsung General Chemicals Co.), and 26 (VS410) wt.% (manufactured by Hyundai Petrochemical Co.). No molecular weight information was provided. The PEMA was evaluated at 0.5, 3 (obtained from Aldrich), and 1 (Polybond 3009 manufactured by Uniroyal) wt.% of maleic anhydride. No additional information about the molecular weight of the PEMA was provided.

The polymers were melt blended first and then Cloisite 15A was added at a 5 wt.% loading in the polymer. The equipment was described as an "internal mixer." No additional information was provided as to the manufacturer or the physical description of the mixer. The mixing conditions were 100°C at 80 r/min for 20 min. No additional processing information was provided. The test samples were compression molded at 100°C for 1 min. The pressure was not provided. The samples were characterized by WAXS. No TEM pictures were provided. Hence, no information was provided to determine the montmorillonite orientation in relation to the direction of stress for the modulus evaluations.

WAXS indicated that higher levels of vinyl acetate and PEMA could provide more efficient dispersion of the montmorillonite. Since no TEM pictures were provided, the degree of exfoliation of the montmorillonite in the polymer blends cannot be assessed definitively. No experimental design was provided and no statistical analysis was carried out. No analysis was provided to determine the statistical relevance of the presumed significant independent variables, vinyl acetate and PEMA content.

One study is provided to quantify the synergistic effects of blending EVA and EMA on the tensile modulus in the presence of montmorillonite. Unfortunately, controls were not evaluated (the tensile modulus of EVA and EVA–EMA polymer blends without the presence of montmorillonite).

The tensile modulus for the EVA blends with 5 wt.% loading of the organomontmorillonite decreased steadily from approximately 1800 kg/cm^3 for the EVA with 3 wt.% vinyl acetate content to approximately 250 kg/cm^3 for the EVA that contained 26 wt.% of vinyl acetate. The addition of 10 wt. % of the PEMA with 1% content of maleic anhydride consistently resulted in a significant increase in tensile modulus. Approximately a 500 kg/cm^3 increase was measured for the 3% vinyl acetate containing polymer to a doubling of the tensile modulus value for the 26% vinyl acetate containing copolymer.

Additional work is needed to ascertain the value of this polymer blend in achieving an exfoliation of organomontmorillonite that results in enhanced mechanical performance.

1. The test samples must be injection molded.
2. TEM must be utilized to determine the degree of dispersion of the montmorillonite and the orientation of the particle in relation to the direction of stress.
3. A proper experimental design must be implemented to determine the significance of the organomontmorillonite concentration, the organomontmorillonite type, and the concentration and type of the EVA and PEMA in the polymer blends.

An illustration of the importance of experimental design in identifying the significant independent variables that relate to the effective dispersion of organomontmorillonite in EVA and the measured dependent variable, Young's modulus, is provided by Tellekeratne *et al.* [10]. The EVA contained 9 wt.% vinyl acetate (manufactured by Cryovac Australia Pty). The molecular weight of the EVA was not provided. The sodium counterioned montmorillonite was provided by Unimin Australia, Ltd. The exchange capacity was not provided. The quaternary ion modification of the montmorillonite was done with an unusual cetyl dimethylethyl ammonium bromide. The performance of this quat on montmorillonite should be similar to the hydrogenated tallow trimethyl quaternary ammonium ion that has been employed in work described above. The amount of quat that was exchanged on the montmorillonite was not provided. All of the polymer–montmorillonite composites were evaluated at 5 wt.%. The processing study was done with a Haake Rheocord 90 internal mixer (manufactured by Instron Corp.). The significant independent processing variables were temperature (90 and 100 °C), rotor r/min (30 and 110), and mixing time (10 and 20 min). The replication of the center point for this two-level design (95 °C, 15 min, and 70 r/min) was employed as an error estimate. Note the error below Table I on page 2653 that records the center point r/min at 170. Also note that because there are three variables evaluated, the three-variable interaction can be utilized as an error estimate. Significant three-variable interactions are seldom found in nature. This work is no exception. The Young's modulus of the EVA without montmorillonite was provided. The test samples were compression-molded. The conditions were not provided. The test samples were conditioned at 25°C and 50% relative humidity. Note that because of the significant sensitivity of vinyl acetate polymers to plasticization by water, the moisture content will also influence the mechanical properties of the composites.

WAXS was employed to measure the degree of exfoliation. The data indicates an intercalated morphology. Environmental scanning electron microscopy (ESEM) was utilized to evaluate the degree of dispersion and particle alignment. The dispersion efficiency of montmorillonite in the EVA in the one picture provided (the processing conditions were not specified) indicated a poor dispersion. TEM pictures were claimed in the text to have been taken. No pictures were provided; no subsequent publications by these authors provided any TEM pictures of these composites. The orientation of the particles in relation to the applied stress was not provided. The relationship to predict the relative changes in the significant independent variables and the dependent variable (Young's modulus) was not linear. The best fit was a second-order quadratic equation. The square of the correlation coefficient (R^2) was 97% (a good fit). The significant variables at or above the 90% confidence level were:

- a positive value for the two-variable interaction of the temperature and the mixing time (a positive synergy between mixing time and temperature);
- a positive value for the mixing time (a negative value is listed in Table III);
- a negative two-variable interaction for temperature and r/min. The contour diagrams indicate that the optimal conditions for the maximum value of the Young's modulus lie outside the design space. When the optimal conditions are evaluated experimentally, the modulus values are consistently higher (approximately 20 MPa) than predicted (approximately 250 to 280 MPa; the modulus for EVA without the montmorillonite was approximately 140 MPa).

The effect of increased shear energy on the dispersion of Cloisite 20A in polyethylene was investigated by Swain and Isayev [11]. The increased shear energy was provided by ultrasound as the polymer–Cloisite 20A melt exited the extruder. The extruder was a single-screw design. This is not the ideal choice based on previous work discussed in the early part of Chapter 6 on nylon 6. The efficiency of the single-screw extruder was enhanced by a Maddock and melt star mixing section. The ultrasound was utilized at the exit of the extruder. Two horns operated at a slit die at the exit of the extruder with a frequency of 20 KHz. The slit die had an opening gap of 4 mm. The amplitudes of the ultrasound were evaluated at 5, 7.5, and 10 μm. The time of the polymer–Cloisite 20A exposure was varied with flowrates (0.25, 0.50, and 0.75 g/s) with a corresponding residence time of 21, 10, and 7 s. The concentrations of Cloisite 20A in the polymer were 2.5, 5.0, and 10.0 wt. %. The high-density polyethylene (HDPE) was HMN 4550–03-Marlex from Chevron-Phillips.

Caution must be exercised when utilizing ultrasound to disperse organomontmorillonite in a polymer melt. The energy of the ultrasound is capable of polymer chain scission. Unfortunately, molecular weights of the polymer as a function of ultrasound energy were not provided.

WAXS and TEM indicated a direct correlation between increased intercalation and exfoliation of Cloisite 20A as a function of increased ultrasound amplitude and residence time. The TEM picture of the ultrasound sample (an extruder feedrate of 0.5 g/s with an amplitude of 5 μm) indicated poor alignment of the montmorillonite. This will compromise the ultimate modulus performance. Optimal results seemed to be achieved at the 2.5% loading of Cloisite 20A.

Unfortunately, the Young's modulus results as a function of ultrasound amplitude are provided only for the 5% loading of Cloisite 20A. The Young's modulus values appear to plateau as a function of ultrasound amplitude (423–425 MPA) and are slightly lower than the unsonicated value of 459 MPa for the 5% Cloisite 20A loaded polyethylene. Pure polymer that was not sonicated has a Young's modulus of 317 MPa.

The largest change in mechanical properties was percent elongation to failure. The values increased from 487% for the unsonicated composite to a range of 835 to 973% for the sonicated composites with a maximum value at approximately 7 μm amplitude for the 2.5 and 5% loaded polymer composites. Pure polymer without sonication has a percent elongation to failure of 908%. It is tempting to suggest that the molecular weight of the polyethylene decreases with increasing amplitude for these composites. Additional work needs to be carried out with this apparently promising process for the preparation of polyethylene–montmorillonite nanocomposites. Improved extruder designs (for example, the utilization of a twin-screw extruder) and protocol should be employed. Ultrasound conditions need to be identified that do not compromise polymer molecular weight. Injection molding procedures need to be implemented to insure montmorillonite alignment in the direction of modulus measurements of the polymer composites.

To further the understanding of the relationships between polar modification of polyethylene and the successful preparation of polyethylene–montmorillonite nanocomposites, a novel modified polyethylene prepared by the reaction of fluorine and oxygen gas with polyethylene was studied by Shah *et al.* [12]. The nature of the polyethylene is modified by 10–15 atom % of fluorine and oxygen. The oxygen modifications result in the development of ketones, aldehydes, and carboxylic functionalities on the polyethylene. The nature of the polyethylene surface has become hydrophilic enough to provide for a stable dispersion in water.

The preparation and characterization of the polymer nanocomposites are identical to that in the work cited above by Hotta and Paul [1]. The controls are HDPE (HiD 9055 manufactured by Chevron Phillips), HDPE modified with fluorine and oxygen (Inhance HD-1800 manufactured by Inhance Fluoro-Seal), and HDPE–g–MA (Fusabond E MB265D manufactured by DuPont, Wilmington, Delaware). Note that the HDPE–g–MA in this study is different (LLDPE–g–MA; Fusabond MB266D) than in the previous work by Hotta and Paul cited above. Cloisite 20A was again employed as the organomontmorillonite for the preparation of the polymer nanocomposites. The percent loading of the montmorillonite varied from approximately 2.25 to 5.2 wt.%.

WAXS and TEM evaluations indicated that the nanocomposites prepared with the HDPE–g–MA demonstrated superior exfoliation when compared to the other polymers. The exfoliation was not as good as in the work above with optimal exfoliation of nylon 6. The relative increase of the modulus for the fluorine and oxygen modified polyethylene was approximately a factor of 2 with a loading of 5.55% montmorillonite; the relative increase for HDPE–g–MA was approximately a factor of 1.7 at 4.97% montmorillonite.

The pure polyethylene modified with fluorine and oxygen is much stiffer than the other polyethylene controls; the modulus at approximately 5.55% montmorillonite loading was significantly higher (about 1.9 GPa) compared with HDPE (0.768 GPa at 5.2% loading) and HDPE–g–MA (approximately 1.4 GPa at 4.97% loading). The authors point out legitimately that an optimization of the modification of the polyethylene with fluorine and oxygen could provide further enhancements of mechanical properties with an increase in a balance of performance.

In situ polymerization of polyethylene in the presence of montmorillonite is a compelling approach to the preparation of exfoliated polyethylene–montmorillonite nanocomposites. The concern in this approach is similar to the strategy employed in the preparation of nylon 6–montmorillonite nanocomposites by *in situ* polymerization pioneered by Toyota as found earlier in Chapter 6. The concern about the use of traditional catalysts for the preparation of polyethylene with montmorillonite is in their sensitivity to protons and bases found at the edges of the montmorillonite and the inherent moisture content associated with the counterions of the montmorillonite.

A more forgiving catalyst, $[(2\text{-}(2\text{-}C_6H_4(i\text{-propyl}))N{=}C(CH_3))_2C_5H_3N]FeCl_2$, was exchanged onto the montmorillonite by Guo *et al.* [13] along with methylaluminoxane (MAO) for the initial formation of polyethylene oligomers. The subsequent addition of MAO and $(CH_3)_2Si(Ind)_2ZrCl_2$ was utilized to complete the polymerization of the polyethylene.

WAXS indicated the possible full exfoliation of the montmorillonite in the polyethylene. No TEM pictures were provided. The tensile strength of the polyethylene increased from approximately 29.8 MPa for the pure polymer to approximately 33.6 MPa for a 2.8 wt.% montmorillonite content in the polyethylene. The disadvantage of this approach is the necessity to utilize solvents for the polyethylene–montmorillonite synthesis.

Work by Lee *et al.* [14] provides a transition from discussions centered on polyethylene–montmorillonite nanocomposites to discussions focused on polypropylene–montmorillonite nanocomposites. An extruder was not employed to prepare the polymer nanocomposites. A Haake Banbury internal mixer prepared the polymer–montmorillonite concentrates (180°C, 10 min, and 50 r/min) and the final polymer composites. The montmorillonite in this study was Kunipia-F from Kunimine, Japan. Octadecylamine served as the organic modification for the montmorillonite. The polymer concentrates were prepared with polypropylene grafted with maleic anhydride (PP–g–MA; Umex 1010 from Sanyo) and PE–g–MA (Umex 2000 from Sanyo). The PP–g–MA-organomontmorillonite concentrates demonstrated superior dispersion of the montmorillonite in relation to the PE–g–MA concentrates. Hence, most of the effort in preparing and characterization of the final polyethylene composites centered on the PP–g–MA-montmorillonite concentrates. The concentrations of montmorillonite in the concentrates were 12.5, 30,

41.7, and 50 wt.%. The PP–g–MA in the final polymer composites was a constant 7 wt.%. The organomontmorillonite in the final composites were 1, 3, 5, and 7 wt.%. Hence, the polyethylene in the final composites were 92, 90, 88, and 86 wt.%. The polyethylene was HDPE PH143 from Hyundai.

WAXS and TEM indicated that intercalated and exfoliated montmorillonite was present in the final composites. The modulus for the composites varied from approximately 520 MPa for the pure polyethylene to 760 MPa for the polyethylene composite with 7 wt.% of organomontmorillonite.

The preparation of PP–montmorillonite nanocomposites is fraught with greater difficulties when compared to that with polyethylene. For example, isotactic PP is not compatible with syndiotactic PP. Knowledge of the nature of the polypropylene that is utilized to prepare the composites is critical to the successful achievement of exfoliation of the montmorillonite. PP requires the utilization of antioxidants in the formulation and compounding of composites. The tertiary carbon on the polymer backbone associated with the pendant methyl group is very susceptible to oxidation and the resultant chain scission.

Kim *et al.* [15] evaluated the significance of a standard antioxidant for PP (Irganox 1010(A); manufactured by Ciba Specialty Chemicals) at a 5 p/h content in PP for the exfoliation of organomontmorillonite in the polymer. The polymer was YUPLENE H236W manufactured by SK Co. The organomontmorillonite was prepared from Kunipia F and protonated octadecylamine exchanged at the exchange capacity of the montmorillonite. The polymer composites were produced with a Brabender with a 30 cc capacity bowl at 180°C. The mixing was at 50 r/min for 5 min. The test samples were prepared by compression molding. The concentrations of the organomontmorillonite in the PP with and without antioxidant were 1, 3, and 5 wt.%. The (001)*d*-spacing of the montmorillonite in PP measured by WAXS without the presence of the antioxidant remained at approximately (001)*d*-spacing of the organomontmorillonite (approximately 2.2 nm). When the antioxidant was present in the PP, the (001)*d*-spacing increased to 3.35 nm for the PP–montmorillonite composite with a 1% loading.

TEM pictures indicated intercalated composites with the montmorillonite oriented in one direction. The antioxidant has weakly acidic phenolic functionalities associated with the molecule. These functionalities provide the potential for exchange at the edges of the organomontmorillonite.

Edge modification of organomontmorillonite is reputed to increase the ability of the polymer to enter the galleries of the organomontmorillonite. This could account for the increased gallery spacing of the organomontmorillonite present in the PP with the antioxidant present.

SAXS was run during the stress–strain evaluations of the samples. The presence of montmorillonite in the PP apparently prevents the change in the order of the PP crystals to the direction of the applied stress. This behavior is consistent with the restrictive influence of montmorillonite in forming a constrained polymer region around the montmorillonite as discussed in Chapter 4 on barrier performance.

The Young's modulus of the composites increase from approximately 0.75 GPa for PP with the antioxidant to approximately 0.85 GPa for a 1% loading of organomontmorillonite. The 3 and 5% loaded composites have about the same values. This is consistent with intercalated organomontmorillonite in the polymer.

Zhang *et al.* [16] evaluated the same organomontmorillonite described above with a functional monomer (MA) and initiator (2,5-dimethyl-2,5-di (*t*-butylperoxy)hexane) associated with the organomontmorillonite surface that will graft to the PP (J-170S manufactured by Honam Petrochemical Co.). The grafting reaction with PP was performed in a Haake internal mixter at 180°C for 20 min. This concentrate contained approximately 30% organomontmorillonite. This master batch was subsequently blended with PP, using the same equipment, at 190°C for 15 min. The final PP–organomontmorillonite composites contained 1, 2, 3, and 4 wt.% organomontmorillonite. The degree of grafting was not determined.

The WAXS and TEM of the 2% montmorillonite containing PP composite indicated an improved montmorillonite dispersion when compared to the above work. Apparent errors are found in the abstract of this work. The WAXS of the MA modified organomontmorillonite indicates a (001)*d*-spacing of 3.2 nm instead of the 30 nm reported; the organomontmorillonite without the MA modification appears to have a (001)*d*-spacing of 1.96 nm, instead of the reported 19.6 nm. This correction is important regarding the proper interpretation of the results presented. The test samples were compression molded. The TEM indicates a random orientation of the montmorillonite. This is unfortunate and compromises the mechanical results presented based on the models presented in Chapter 5 that predict the mechanical behavior of polymer–montmorillonite nanocomposites.

Additional work (similar to the characterization found in Chapter 5 for nylon 6) is necessary to fully characterize the degree of exfoliation of montmorillonite; the montmorillonite needs to be oriented in relation to the direction of stress for proper characterization. DMA indicates reinforcement of montmorillonite in the PP as a function of montmorillonite concentration. A careful study of this approach is expected to result in significant improvements in mechanical properties.

Chung *et al.* [17] adopted an optional approach of functionalizing the montmorillonite (Cloisite Na; Southern Clay Products) by exchanging a diammonium ion (1,12-diaminododecane neutralized with HCl) onto the montmorillonite surface. MA-grafted PP (MAPP) was dispersed with the organomontmorillonite in a solvent (1,2,4-trichlorobenzene) and reacted with titanium butoxide catalyst at 160°C to produce amide linkages between the diammonium ions on the surface of the montmorillonite and the anhydride functionality on the PP. The polymer composite was precipitated from the solvent with methanol and dried. These master batches contained 5, 10, and 20 wt.% of organomontmorillonite. Pure PP was compounded with these master batches in a Brabender at 200°C and 50 r/min for 10 min. Test samples were prepared by compression molding at 200°C. Unfortunately, the type and manufacturer of the PP and MAPP were not identified.

The WAXS and TEM indicated poor levels for exfoliation and dispersion. The tensile modulus increased from approximately 470 MPa for the blend of 75 wt.% of PP with 25 wt.% of MAPP without montmorillonite to 640 GPa for the blend with 25% MAPP containing 5% montmorillonite. The blends with MAPP containing the higher concentrations of montmorillonite provided successively lower tensile modulus values.

The results from this approach are not very encouraging. Having a multifunctional surface treatment that exchanges onto the surface of montmorillonite apparently compromises exfoliation efficiency, presumably through the possibility of exchanging onto more than one montmorillonite surface and binding them together.

A superior strategy to the above approach is provided by Cui and Paul [18]. The quality of this work is enhanced by the full description of the polymers that were utilized and the complete characterization of the composites that were prepared. The PP that was employed in this study was Pro-Fax PH020 manufactured by Basell. MAPP (PP–g–MA) was PB3200 provided by Cromption with a MA content of 1.0 wt.%. The diamine and montmorillonite (Cloisite Na) were identical to the ones utilized in the above study. Cloisite 20A was utilized as a control organomontmorillonite. This is a superior choice to the organomontmorillonite employed in the studies above. The composites were prepared with a DSM Micro 5 compounder. The test samples were prepared with a DSM microinjection molder. This is a superior protocol in relation to the compression-molded test samples prepared above. PP–g–MA was reacted directly with the diamine in a Brabender at 195°C and 50 r/min for 5.5 min. The amine functionalities on the PP–g–MA (PP–g–NH_2) were converted to ammonium ions by refluxing a toluene solution of the polymer with HCl at 70°C for 7 h. The polymer–montmorillonite composites were prepared with a montmorillonite loading of 1, 3, 5, and 7wt.%. The optimal ratio of functionalized PP to Cloisite Na and Cloisite 20A was determined to be 1.

WAXS and TEM indicated that PP–g–MA blended with Cloisite 20A provided the best dispersion and exfoliation of the montmorillonite in the PP/functionalized PP/montmorillonite blends at 5% loading of the montmorillonite. The degree of exfoliation was good but inferior to the excellent exfoliation achieved with nylon 6 that was discussed here in Chapter 5.

These observations are reflected in the superior modulus values. The modulus for the pure polymer was 1.49 GPa.; the modulus value for the montmorillonite composite at 5% loading was 2.14 GPa. Composites with Cloisite Na did not vary in modulus significantly with montmorillonite loading or polymer blend composition. The protonated amine functional PP (PP–g–NH_3^+) was more effective than the unprotonated amine functional PP in achieving dispersion and exfoliation of the montmorillonite in PP and not as effective as PP–g–MA.

The difficulties in adopting PP–g–MA as the compatibilizer for exfoliation of montmorillonite into PP are the increased cost associated with the MA modification and the compromise in molecular weight of PP upon modification. The decrease in molecular weight will result in a decrease in mechanical properties.

An evaluation of the significance of processing and MA concentration (and consequently, molecular weight) of the PP–g–MA compatibilizer was carried out by Wang *et al.* [19]. The processing variables were evaluated with an intermeshing, corotating, self-wiping, twin-screw extruder with a length: diameter (*l*:*d*) ratio of 45 (PSM30 manufactured by Sino-Alloy Machinery Inc.). This equipment will provide increased dispersive and distributive mixing with an increase in residence time when compared to the DSM equipment above. The test samples were prepared by injection molding. Isotactic PP (Yungsox 1040) was provided by Yung-Chia Chemical Co. Four different PP–g–MA polymers were evaluated with MA contents of 0.5 (PB 3150), 1.0 (PB 3200; evaluated in the above study), 1.2 (PB 3000), and 4.0 (E 43) wt.%. The PB materials were obtained from Crompton and the Epolene E 43 was produced by Eastman Chemical Co. Evernox-10 (Everspring Chemical Co.) at a 0.2 wt.% was added to all the composites as a heat stabilizer.

The organomontmorillonite was prepared by exchange with octadecylamine (I.30 P manufactured by Nanocor).

The montmorillonite in I.30 P is significantly different from Kunipia F that was used in the above studies with octadecylamine modification. The extruder is divided into 10 equal sections (barrels). The temperature of the extruder was varied from 165°C in the second barrel to 180°C in the fifth barrel. The screw speed (200 r/min) was held constant for all of the preparations. Two processing variables were adopted:

1. Load the PP in a hopper to feed into the extruder and then feed the PP–g–MA-organomontmorillonite mixture at barrel number 1 onto the PP.
2. Feed the PP–g–MA-organomontmorillonite into the hopper and add the PP at barrel 4 (referred to as side feeding).

The advantage of the side feeding is that the PP–g–MA-organomontmorillonite experiences dispersive mixing in barrels 1 and 2 with subsequent distributive mixing before the PP is added.

WAXS and TEM indicate that, with all of the composites, side feeding produces superior montmorillonite dispersions. The PP–g–MA compatibilizer with the lower MA content and higher molecular weight (PB 3150 and PB 3200) provided the best dispersion.

Intercalation of the montmorillonite galleries with PP without full exfoliation was observed. Caution must be exercised when evaluating the tensile modulus results. The modulus of the polymer blends needs to be compared with and without montmorillonite present (as in the work above) to obtain a more valid comparison of the impact of the different montmorillonite composites on mechanical properties. PB 3200 would appear to be the best choice (as in the work above) for a PP–g–MA compatibilizer for the preparation of a PP–montmorillonite composite. The tensile modulus for a 5 wt.% organomontmorillite containing PP composite with 15 wt.% PB 3200 as a compatibilizer increased from 936 MPa with the hopper feed to 1037 MPa for the side-feeding protocol. The pure PP had a tensile modulus of 690 MPa.

An examination of the significance of dispersion protocol is provided by Leuteritz *et al.* [20]. The PP and PP–g–MA was dispersed with the organomontmorillonite with sequential utilization of a Ko-Kneader (BUSS MKS 30) and twin-screw extruder (ZE 25 UTS) to finish the dispersion. The processing temperatures varied between 200 and 220°C at 300 r/min. The final composite composition contained 7.5 wt.% PP–g–MA and 5 wt.% montmorillonite. The PP–g–MA was reported to have 0.5% MA content from Crompton. It is assumed that this PP–g–MA was PB3150, which was utilized in the study above. The PP was Moplen HF500N (manufactured by Basell). The organomontmorillonite was similar to Cloisite 20A. The montmorillonite had a lower exchange capacity (75 meq/100g of montmorillonite). Hence, the organic content of the dimethyldistearyl quaternary ammonium exchanged onto the montmorillonite was lower than Cloisite 20A. The test samples were prepared by injection molding. The modulus was improved by 148% when compared to the pure PP.

The results of the above study with PB3150 and octadecylamine-exchanged montmorillonite indicated a tensile modulus improvement of 138.6% with the hopper feed and 145.4% improvement with the side-feed protocol. Further work with Cloisite 20A and the side-feed protocol of the above work at the lower PB3150 content in this study could produce further improvements of montmorillonite dispersion and subsequent improvements in the mechanical performance of PP.

An additional feature of this work was the evaluation of the significance of silane modification of the organomontmorillonite (3-aminopropyltrimethoxysilane, 3(2-aminoethyl)aminopropyltrimethoxysilane, and 3(6-aminohexyl)aminopropyltrimethoxysilane) with regard to improved montmorillonite dispersion and the mechanical properties of PP–montmorillonite polymer composites.

No significant advantages were demonstrated with any silane modification. SAXS was employed to measure the (001)*d*-spacing of the organomontmorillonite and the montmorillonite in the PP composite. This is not the best choice; WAXS provides more satisfactory results. Indications are that the (001)*d*-spacing of the organomontmorillonite is not significantly altered by silane modification. TEM was more enlightening. Dispersion and intercalation of the polymer into the galleries of the organomontmorillonite was superior without silane modification.

The importance of utilizing injection molding processing to prepare polymer–montmorillonite composites for testing has been emphasized throughout this chapter. An illustration of the importance of this process for the preparation of PP–montmorillonite composites is found in the work of Wang *et al.* [21]. The polymer composites were prepared in the mold without the use of extruder processing before molding. Isotactic PP that was evaluated in this work was manufactured by WuHan Petroleum Chemical Co. The PP–g–MA compatibilizer contained 0.9 wt.% MA (manufactured by ChenGuang Co.).

The organomontmorillonite was similar to Cloisite 20A; the montmorillonite had a lower exchange capacity (68.8 meq/100g of montmorillonite; received from RenShou).

The PP–montmorillonite composites were prepared by two processes:

- injecting the PP–montmorillonite composite melts at 190°C directly into the mold and allowing the melt to cool to room temperature before testing;
- cycling the composite melts at 1.0 Hz frequency through the mold until cooled to a solid.
- The residence time was approximately 10 min, pressure in the mold was 4 MPa, and the injection pressure was 90 MPa. Three composites were prepared at each condition:
- PP–pure montmorillonite at 5 wt.% loading;
- PP–organomontmorillonite at 5 wt.% loading;
- PP–PP–g–MA-organomontmorillonite with a PP–g–MA concentration of 10 wt.% and an organomontmorillonite loading of 5 wt.%.

WAXS and TEM indicated that the cycling produced significant improvements in montmorillonite dispersion and intercalation of the PP into the montmorillonite galleries. The formulation with the PP–g–MA compatibilizer demonstrated the best dispersion. The composite prepared without the compatibilizer and the organomontmorillonite was remarkably good.

Unfortunately, notched Izod impact strength was the only mechanical property measured in the work. Impact strength based on notched Izod is always problematical. A small defect in the sample can lead to a major failure in the sample during testing. The notched Izod impact strength of the pure PP increased from approximately 2.9 kJ/m^2 for the uncycled process to 4.6 kJ/m^2 for the cycled process. The cycling process was significant in aligning the polymer chains for improved mechanical performance. The notched Izod impact strength for the composites prepared with PP and organomontmorillonite without PP–g–MA compatibilizer increased from approximately 3.2 kJ/m^2 for the uncyclized process to approximately 15.3 kJ/m^2 for the cyclized process! When the compatibilizer was present, notched Izod impact strength for the composites increased from approximately 3.3 kJ/m^2 for the uncyclized process to approximately 13.6 kJ/m^2 for the cyclized process. As expected, the addition of the lower molecular weight compatibilizer lowered the toughness of the composite.

One can speculate on the improvement of PP–organomontmorillonite nanocomposite mechanical properties without a compatibilizer if an effective twin-screw extrusion process with a side feed for the organomontmorillonite similar to that described above is integrated in to an injection mold with a gate that provides a similar shear history to that generated by the cycling in the mold of this study.

An expanded process research study of the extrusion to produce PP–organomontmorillonite polymer composites is provided by Modesti *et al.* [22]. The significant independent variables were:

- the temperature of the extruder (high: 200–210°C and low: 170–180°C);
- r/min (250 and 350);
- compatibilizer content (0 and 6 wt.% PP–g–MA; Fusabond MD551D with 1% MA content manufactured by DuPont);
- organomontmorillonite content (0, 3.5, and 5 wt.%).

PP (PPH 7062) was obtained from Total Petrochemicals. The extruder was an intermeshing, corotating twin-screw extruder with a L:D of 40. The extruder manufacturer was not described. The organomontmorillonite was similar to Cloisite 20A (Dellite 72T manufactured by Laviosa Chimica Mineraria). The test samples were prepared by compression molding.

WAXS and TEM analysis of the composites indicate that the extruder conditions that produced the best organomontmorillonite dispersion were a low temperature and higher screw speed. The dispersion of organomontmorillonite into this polymer system responds favorably to higher shear conditions (higher melt viscosity and higher r/min). The addition of the PP–g–MA compatibilizer to the PP was beneficial to improving the dispersion of the organomontmorillonite.

Full exfoliation of the organomontmorillonite was not observed; the best dispersion appeared to be principally intercalated polymer in the montmorillonite galleries. The tensile modulus values for the polymer composites were commensurate with the observed morphology differences in the composites. The best performer was the composite prepared at a low extruder temperature, high r/min, containing the compatibilizer, and with a 5% loading of organomontmorillonite. The tensile modulus for this composite was 130% greater than the pure PP. One must be cautious when comparing these results with test samples prepared by injection molding found above.

Polymer blends of PP have been employed to expand the commercial applicability of PP for engineering applications. Key parameters of the blends that relate to mechanical performance are the particle size and particle size distribution of the polymer-dispersed phase. The product designation for these blends is sometimes referred to as thermoplastic polyolefin (TPO). Paul *et al.* [23–26] published an excellent series of four articles that identifies the significant structure–property relationships of TPO–montmorillonite nanocomposites that relate to mechanical properties. The processing equipment is identical to that employed in the earlier work on nylon 6–montmorillonite nanocomposites. A Hakke corotating, twin-screw extruder (l:d = 10) was utilized to prepare TPO dispersions from PP and ethylene–octene copolymer by melt blending (170 °C at 280 r/min). Ethylene–octene copolymer is a popular rubber modifier for commercial TPO. The rubber modifications were 10, 20, 30, and 40 wt.%. The manufacturers of the PP and ethylene–octene copolymer were not identified. A master batch of a 1:1 mixture of PP–g–MA (1.0 wt.% MA) and organomontmorillonite (it appears to be Cloisite 20A based on the quat identification and (001)*d*-spacing by WAXS, although it was not identified) was prepared by the same protocol. The montmorillonite content of the master batch was 30 wt.%. The final PP–elastomer–organomontmorillonite dispersions were prepared by melt blending the TPOs with the master batch at ratios to prepare approximately 1, 3, 5, and 7% montmorillonite content. The test samples were prepared by injection molding. The protocol for the determination of the degree of intercalation and exfoliation of the montmorillonite in the TPO composites was identical to that for the characterization of the nylon 6 nanocomposites described in Chapter 5.

WAXS and TEM indicated a mixture of exfoliated and intercalated montmorillonite in the master batch. The degree of exfoliation was compromised at the higher montmorillonite content (the number-average aspect ratio decreased from 63 for 1% loading to 37 for 3% loading and 31 for 7% loading). The observations for these particle-size determinations were made parallel to the tangential perspective of the flow direction of the injection molder for sample preparation. The significance of this will be discussed in the following section.

AFM (Digital Instruments Dimension 3100 with a Nanoscope IV controller) was employed to evaluate the particle size of the rubber-dispersed phase as a function of montmorillonite loading in the composite, because the higher temperatures experienced by TEM measurements will distort the softer dispersed phase. The particle size of the rubber-dispersed phase decreased significantly as a function of montmorillonite loading. For example, at a 30 wt.% loading of the rubber-dispersed phase, the number-average particle size decreased from 1.88 μm without montmorillonite present in the composite to 1.18 μm with a 1% loading of montmorillonite, 0.69 μm for 3% loading, and 0.44 μm for 7% loading. The number-average aspect ratio of the rubber dispersed phase changed from 1.5 without the presence of montmorillonite to 2.8 for the 7% loading of montmorillonite in the TPO–montmorillonite composite.

The observations for these measurements and those of the flow direction of the injection molder that prepared the test samples were made in parallel. The significance of this is discussed in the following section.

This dramatic change in morphology for the 30% rubber containing TPO–montmorillonite composite resulted in a substantial increase in impact resistance (notch Izod; TMI Izod tester with a 6.8 J hammer and 3.5 m/s impact velocity) from approximately 100 J/m for the TPO without montmorillonite and 1% loading of montmorillonite to greater than 600 J/m for the TPO with 3% loading! Increases in montmorillonite content to 5 and 7% in the TPO composite did not decrease the impact resistance, as expected, with conventional dispersed phase-reinforcing additives. The impact resistance continued to increase with montmorillonite loading! The TPO with a 40% rubber content demonstrated a similar behavior but not as dramatic. The TPOs with the lower rubber content (10 and 20%) did not indicate this dramatic transition as a function of montmorillonite loading.

This transition in polymer composites containing rubber-dispersed phases is sometimes referred to as the brittle–ductile transition (for obvious reasons). The influence of montmorillonite on this transition through the alteration of the rubber-phase morphology for this TPO is very significant. The tensile modulus for this TPO with 30% rubber loading as a function of montmorillonite content increased from approximately 0.8 GPa with no montmorillonite content to approximately 1.5 GPa for 7% montmorillonite content. It is difficult to deconvolute the contribution of the montmorillonite from the contribution of the rubber-dispersed phase to these mechanical properties.

The second publication [24] examines the morphology of each dispersed phase as a function of changes in concentration of both dispersed phases in the TPO composite. Examination of the end and side of the test samples prepared by injection molding indicates anisotropic particle size and aspect ratios as functions of orientation. For a 30% rubber concentration in TPO, the rubber-dispersed phases appear to be larger (number-average diameter = 1.52 μm) when viewed at the edge of the test samples as opposed to the end of the test sample (number-average diameter = 1.18 μm) at 1% montmorillonite loading. The number-average aspect ratios of the rubber phases are similar (1.4 for the edge view versus 1.8 for the end view). At 7% montmorillonite loading, the number-average diameters are significantly reduced and are similar for both views (the edge view is 0.45 μm and the end view is 0.44 μm). The aspect ratios at 7% montmorillonite loading are much larger (4.2 for the edge view and 2.8 for the end view).

These dramatic changes in the rubber-phase morphology resulted in significant differences in the CTE as a function of montmorillonite loading and orientation of the test samples. The changes in particle size of the montmorillonite as a function of end versus edge views seem to support a proposed mechanism of the exfoliation of montmorillonite from an intercalated collection of particles. The mechanism describes the exfoliation of montmorillonite as a process where individual particles are separated from the collection by peeling away separate particles starting at their edges. When the intercalated particles are viewed from the end of the test sample, the individual particles are seen to peel away and the particle lengths and thicknesses are smaller as a function of increasing montmorillonite concentration in relation to the edge views, which consist of observing only the intercalated particles. For the master batch of PP–g–MA-montmorillonite, the number-average particle length for the edge view decreased from 288 nm for 1% montmorillonite loading to 225 nm for 7% montmorillonite loading and the number-average particle length for the end view varied from 271 nm to 183 nm for the same concentration range of montmorillonite. The number-average particle thickness for the edge view increased from 4.6 nm to 7.3 nm over this concentration range and decreased from 6.4 nm to 4.9 nm for the end view over the concentration range.

The direction of the flow of the injection molder had a significant influence on the anisotropic morphology of the rubber and montmorillonite-dispersed phases. Synergy between the concentration of the rubber phase and montmorillonite during injection molding provided the anisotropic morphology in the preferred orientation for improved mechanical properties. This thesis is supported by the significant decrease in the CTE (from about 6.1×10^{-5} mm/mm°C for pure PP to about 5×10^{-5} mm/mm°C for the TPO composite with 40% rubber and 7% organomontmorillonite) of the test samples in the flow direction at the higher rubber (30 and 40%) and higher montmorillonite (5 and 7%) concentrations. The CTE was measured with a Perkin–Elmer thermomechanical analyzer (TMA 7) over a temperature range of 0 to 30°C.

In the third article of this series, a commercial TPO, CA387 manufactured by Basell, was evaluated as a function of PP–g–MA (Polybond

3200 manufactured by Chemtura with a MA content of 1.0 wt.% MA) and Cloisite 20A by the same protocol described above. The elastomer ethylene–octane content of the TPO was approximately 25 wt.%. Examination of the commercial TPO versus the TPO prepared above by melt extrusion from PP and the elastomer indicated a much higher notched Izod performance (about 600 J/m for the commercial TPO versus about 100 J/m for the melt-blended TPO prepared above). Apparently, the rubber content and particle size of the rubber have been optimized in the commercial TPO.

WAXS and TEM indicated intercalated montmorillonite particles. Evaluations of the particle size of the montmorillonite from the perspective of the edge of the test samples orthogonal to the flow direction indicated a decrease in number-average particle length (623 nm to 143 nm) as a function of increasing PP–g–MA content at a constant 5% Cloisite 20A content (the ratio of PP–g–MA to Cloisite 20A was increased from 0 to 2.0). The number-average particle thickness decreased over this same range from 84.0 nm to 4.1 nm. The number-average of the particle aspect ratios increased from 9 to 43 over this range. PP–g–MA concentration is an important significant variable for improving the dispersion of Cloisite 20A in this TPO–montmorillonite composite.

For this same set of conditions, AFM indicated that the number-average particle size of the rubber-dispersed phase decreased from 0.78 μm to 0.51 μm as the amount of PP–g–MA content was increased. The number average of the aspect ratios of the rubber-dispersed phase increased from 2.4 to 4.42 over this change of PP–g–MA content. The rubber-dispersed phase is distorting into an elongated morphology in the direction of the flow in the injection molder and the particle size is getting smaller as a function of added PP–g–MA.

When the ratio of PP–g–MA to Cloisite 20A is fixed at 1, the number-average particle size of the elastomer significantly decreases as a function of increasing montmorillonite content (0.86 μm for 0% montmorillonite content to 0.55 μm for 7% montmorillonite content). The number-average of the aspect ratios increased from 2.12 for 0% montmorillonite to 4.75 for 7% montmorillonite content in the TPO–montmorillonite composite with PP–g–MA as the compatibilizer. The tensile modulus for the 5 and 7% montmorillonite containing TPO composites increased rapidly up to a PP–g–MA/montmorillonite ratio of 1. The composite of TPO with and without PP–g–MA with no montmorillonite content had a tensile modulus of approximately 0.78 GPa.

When Cloisite 20A was included in the composite, the tensile modulus increased to 1.52 GPa for a ratio of PP–g–MA to montmorillonite of 1 for a 7% loading of montmorillonite and 1.65 GPa for a ratio of PP–g–MA to Cloisite 20A of 2. The notched Izod impact strength increased slightly from approximately 600 J/m for the pure TPO to approximately 650 J/m at 1% montmorillonite content for all of the PP–g–MA: montmorillonite ratios. The impact strength of the composites decreased at higher montmorillonite content for the PP–g–MA: montmorillonite ratios.

A comparison of this work with the above investigations with laboratory-prepared TPO indicates that the addition of organomontmorillonite with a compatibilizer is not a panacea for increasing the impact strength of rubber-reinforced PP. The commercial TPO preparation from Basell provides an optimal toughness that is compromised at higher concentrations of montmorillonite.

The fourth article in this series is an extension of the above work with the TPO, CA 387, manufactured by Basell. The compatibilizer (MA content of 1.0%) and organomontmorillonite are not specified as regards manufacturer and product name. Based on the discussion in this article, the compatibilizer is assumed to be Polybond 3200, manufactured by Chemtura, and the organomontmorillonite is Cloisite 20A.

The morphology of the composites were carefully analyzed by WAXS and TEM for the montmorillonite and AFM for the elastomer-dispersed phase. The additional mechanical property that was measured in this work was the CTE with the same equipment, TMA 7, and protocol that was employed in the second article in this series (see above).

As a function of the anisotropic orientation of aspect ratio for the montmorillonite and elastomer in the preferred direction of the flow of the injection molder, the expansion coefficient of the composite in the flow direction decreased from approximately 9.7×10^{-5} mm/mm °C for 0% montmorillonite content to approximately 5.3×10^{-5} mm/mm °C for 7% montmorillonite content when the ratio of compatibilizer to montmorillonite was 2. Similar values were observed for compatibilizer montmorillonite ratios of 0.5 and 1.0.

As a measurement of the significance of the anisotropic orientation of the dispersed phases (elastomer and montmorillonite), the CTE in the direction orthogonal to the flow direction of the injection molder decreased from approximately 14.4×10^{-5} mm/mm °C for 0% montmorillonite content to approximately 10.5×10^{-5} mm/mm °C for 7% montmorillonite content in the composite for compatibilizer/montmorillonite ratios of 0.5, 1.0, and 2.0. This perspective is consistent with the previous work in this series (see above). Full exfoliation of montmorillonite in these composites would be expected to provide a significantly greater decrease in the CTE.

7.2 Difficulties associated with the preparation of polyimide–clay nanocomposites

The preparation of polyimide–montmorillonite nanocomposites presents challenges that are unique compared with the constraints associated with the polymers discussed above. Polyimides are utilized in the aerospace industry because of their balance of mechanical properties and high thermal stability. Polyimides have found value in the electronic packaging and insulation industrial because of the additional quality of electrical insulation.

The commercial focus in these market areas is on aromatic polyimides. These are prepared in two steps:

- the reaction of a dianhydride with a diamine at room temperature in an aprotic solvent to produce a poly(amide acid);
- increasing the temperature to at least 300°C to remove the solvent and form the polyimide.

The high temperature of the second step precludes the utilization of montmorillonite exchanged with the quaternary ammonium ions described in the work above. These quats are not stable at these temperatures.

A series of articles by Delozier *et al.* [27–30] definitively demonstrates the inappropriate choice of quats utilized in the studies above, owing to their thermal instability at the temperatures employed in the second step of the polyimide synthesis. Organic modification of montmorillonite was explored to provide greater thermal stability and to be commensurate with the chemistry associated with the polyimide synthesis.

In the second article of this series, the ammonium salt of 1,3-bis(3-aminophenoxy)benzene (APB; provided by Mitsui Petrochemical Ind. Ltd) was exchanged onto an excellent dispersion of Cloisite Na in water. Unfortunately, the precise organic content of this organomontmorillonite was not determined (for example, by thermal analysis). As an additional independent variable, the exchange capacity of the montmorillonite was lowered by first exchanging the montmorillonite with lithium as a counterion and by subsequent thermal treatment in air (120, 130,

140, 160, and 170°C). Each montmorillonite sample was held at the prescribed temperature for 24 h. Note that the designation for the cation exchange capacity (CEC) listed in Table 2 on page 2234 of reference [28] should probably be MCM-160-Li instead of MCM-150-Li. The exchange capacity decreases as a function of temperature from 0.71 meq/g of montmorillonite at 120°C to 0.49 meq/g of montmorillonite at 160°C.

The thesis of this investigation is to provide a spectrum of organo-montmorillonites with different hydrophilic–hydrophobic balances by controlling the amount of ammonium ion exchanged onto the montmorillonite surface by steadily decreasing the CEC. Unfortunately, the amount of organic ammonium ion exchanged onto the montmorillonites was not experimentally measured. The organomontmorillonites were dispersed into 1-methyl-2-pyrrolidinone (NMP) with sonication. The poly(amide acid) was prepared by reacting the organomontmorillonite dispersions with an NMP solution of 3,3′, 4,4′-biphenyltetracarboxylic anhydride (BPDA; provided by Chriskev Co., Inc.) and an additional APB–NMP solution at room temperature for 24 h. The solution of poly(amide acid)–organomontmorillonite was drawn down on clean, dry glass plates, dried to a tackfree condition at room temperature with a dry air flow, and cured to the polyimide by heating at 100, 200, and 300°C for 1 h at each temperature. The films were approximately 30 μm thick. The organomontmorillonite content in the polyimide was evaluated at 3, 5, and 8%.

WAXS and TEM evaluations of the composites indicated that the organomontmorillonite with a CEC of 0.7 meq/g produced the best dispersion of the group. The dispersion was good. The organomontmorillonite appeared to be intercalated and randomly dispersed in the polymer. The apparent high viscosity of the dispersion could have interfered with the alignment of the montmorillonite upon draw down. The modulus increased as a function of montmorillonite content from 511 ksi (Kip/in^2 $\approx$ 6.89×10^6 Pa; Kip $\approx$ 4.45×10^3 N) for pure polymer to 681 ksi for 8% organomontmorillonite measured orthogonally to the direction of the draw down and 592 ksi in the direction of the draw down. The mechanical measurements indicate an anisotropic distribution and orientation of the montmorillonite particles in the polymer. Improved shear conditions (similar to those for injection molding of polymer melts evaluated above) that would improve the orientation, distribution, and possibly the degree of exfoliation of the organomontmorillonite in relation to the direction of mechanical analysis would be expected to improve mechanical performance.

The third publication in this series evaluates the role of a diamine with a more flexible 3-oxapentane link between the aromatic amine functionalities (1,5-bis(3-aminophenoxy)-3-oxapentane; BADO) exchanged onto the reduced CEC montmorillonite (0.71, 0.70, 0.63 meq/g of montmorillonite) as a hydrochloride. APB and BPDA were employed with the organomontmorillonites as found in reference [28] to produce the polyimide–organomontmorillonite composites. Thin films of the polyimide–organomontmorillonite composites (approximately 30 μm) for mechanical testing were prepared using the identical protocol described above. The organomontmorillonite loadings in the polyimide were 3, 5, and 8%.

WAXS and TEM evaluations indicate an intercalated, randomly oriented good dispersion of the organomontmorillonite with a CEC of 0.7 meq/g of montmorillonite in the polyimide. The orientations of the TEM pictures were not specified in relation to the direction of the draw down of the poly(amide acid) before cure to form the polyimide. The mechanical properties of the polyimide–montmorillonite composites were evaluated orthogonal to the draw-down direction. The modulus values of the best performing polyimide–organomontmorillonite (BADO exchanged onto the montmorillonite with a CEC of 0.7 meq/g of montmorillonite) as a function of montmorillonite content were virtually identical to the results in the study above (approximately 510 ksi for the pure polymer to approximately 680 ksi for the 8% loading). No performance advantages were observed from exchanging BADO onto the montmorillonite when compared with the APB-exchanged montmorillonite in the previous work. As with the latter, improved mechanical performance is anticipated if shear conditions are adopted that orient the particles in the direction of mechanical testing. Additional shear could also improve the degree of exfoliation of the montmorillonite that would improve mechanical performance.

In the final publication of this series, a polyetherimide–organomontmorillonite composite was prepared that was suitable for melt blending and extrusion. APB was exchanged onto the reduced charged montmorillonite (0.70 meq/g of montmorillonite) and was employed at a 3% loading level in the preparation of the polyimide. The polyimide was different from that studied above (*m*-1,3-phenylenediamine (*m*-PDA) was reacted with [(1-methylethylidene)bis(1,4-phenyleneoxy)] bis-1,3-isobenzofurandione (BPADA)). This polyetherimide is suitable for melt blending. Melt blending was accomplished by melting the polyetherimide control and the polyetherimide–montmorillonite composite in powdered form with a Brabender at 20 r/min at 325°C for 90 min. The melts were then cooled and pelletized. The pellets were fed into a Brabender PL2000 Plasti-Corder with an attached extruder with an L:D of 4.

The WAXS and TEM evaluations indicated an intercalated montmorillonite dispersion. The melt-blending step produced the best orientation of the montmorillonite. Extrusion resulted in a return to a random distribution of the intercalated montmorillonite particles. The increase in tensile modulus was modest (the authors now adopt GPa units instead of ksi). The tensile modulus of the melt-blended composite was higher (2.7 GPa for the pure polymer increased to 3.0 GPa for the 3% montmorillonite). The extruded polymer composite was evaluated as fibers. No mechanical performance advantages were measured for the extruded composite (2.1 GPa for the pure polymer increased to 2.4 GPa for the composite).

Improved extrusion conditions are expected to provide enhancements of montmorillonite alignment in the polyetherimide and possibly improved dispersion and exfoliation efficiency with subsequent improvement in mechanical properties.

7.3 The conundrum of polystyrene–clay nanocomposites

Full exfoliation of montmorillonite in polystyrene has not been achieved. Hence, the full realization of the reinforcing benefit of montmorillonite with regard to the mechanical property enhancement of polystyrene has not been exploited. The definitive discussion that relates to the complex and confounding problem relating to the inability to exfoliate montmorillonite into polystyrene is found in the work of Stretz and Paul [31]. The factors that relate to improved dispersion and exfoliation of montmorillonite in polyolefins that are discussed above do not provide a satisfying response with polystyrene. The conclusion found in this work is the result of extensive investigations of acrylonitrile (SAN) and maleic anhydride (SMA) modified polystyrene [32,33,34].

The first study [31] surveyed a wide range of different quaternary ammonium ion modifications of montmorillonite with Tyril 100 (manufactured by Dow Chemical Co.). Tyril 100 is polystyrene with approximately 25 wt.% acrylonitrile and a weight-average molecular weight of approximately 152000 g/mol. Nine organomontmorillonites were evaluated in this study:

- Cloisite 30B,
- Cloisite 25A;
- Cloisite 20A;
- Cloisite 15A;
- Cloisite 6A;
- montmorillonite exchanged at 95 MER with the dimethyl hydrogenated tallow ammonium ion;
- trimethyl octadecyl quaternary ammonium ion produced from palm oil;
- benzyl dimethyl hydrogenated tallow quaternary ammonium ion;
- bis(hydroxyethyl) methyl dodecyl quaternary ammonium ion produced from coconut oil.

The polymer–montmorillonite composites were produced with the same Haake twin-screw extruder and protocol that was successfully employed with the nylon 6 work described in Chapter 5. The test samples were prepared by injection molding and characterized by the same equipment and protocol described in the successful nylon 6 preparation of polymer–montmorillonite nanocomposites described in Chapter 5. The montmorillonite content in the composites was prepared at approximately 2, 4, and 5.5 wt.%.

WAXS and TEM indicated that the best dispersions were prepared with the trimethyl octadecyl quaternary ammonium and dimethyl octadecyl ammonium ion exchanged montmorillonite. The number-average particle length varied from 172 to 240 nm; the number-average particle thickness varied from 4.19 to 4.5 nm. The Young's modulus increased from approximately 3.5 GPa for the pure polymer to approximately 5.5 GPa for an approximate 5.5% loading of the ammonium ion-exchanged montmorillonite in the SAN.

The Halpin–Tsai prediction for the modulus at 2 wt.% loading of fully exfoliated montmorillonite with an aspect ratio of 100 in this polymer is duplicated by the ammonium ion-exchanged montmorillonite that has an experimentally determined aspect ratio of approximately 53.3. The enhanced reinforcing efficiency of this organomontmorillonite in SAN as a function of aspect ratio was not anticipated.

The second article in this series [32] extrapolates the knowledge of SAN–montmorillonite nanocomposites developed above to a rubber-dispersed polymer (acrylonitrile–butadiene–styrene; ABS). The same equipment and protocol that was employed above was implemented in this work. The ABS was produced by GE (Cyclac GMP 5500 natural). The butadiene is the rubber-dispersed phase. The particle size dispersed phase is based broadly on the TEM. The average particle size appears to be in the range of 200 nm. The ABS was evaluated as stained (exposed to OsO_4 vapors for 24 h) and unstained.

The unstained TEM was the most insightful. The organomontmorillonite selection for this work was more limited in relation to the work above. Cloisite 30B, 25A, 20A, and montmorillonite exchanged with benzyl dimethyl hydrogenated tallow quaternary ammonium ion were evaluated. SAN from the above work was the control. The TEM of the unstained ABS indicates the close association of the Cloisite 30B to the interface of the rubber-dispersed phase particles.

Quaternary ammonium ions exhibit similar behavior as phase transfer catalysts in organic synthesis. Similar effects are expected with organomontmorillonite. The tensile modulus of Cloisite 25A evaluated at 1 and 3% loading in ABS (2.2 GPa for pure polymer increased to approximately 3.0 GPa for 3% loading) and 2, 5.5, and 6.5% loading in SAN (approximately 3.5 GPa for pure polymer increased to 5.7 GPa for 6.5% loading) was superior to the other organomontmorillonites. The exfoliation efficiency of organomontmorillonite in ABS when compared to SAN seemed to be compromised (approximately 10%) by their association at the rubber-dispersed phase.

The third article in this series [33] evaluated the exfoliation efficiency of one organomontmorillonite (montmorillonite exchanged at 95 MER with trimethyl octadecyl quaternary ammonium ion) as a function of acrylonitrile content of SAN. Polystyrene was the control (Styron 685 D produced by Dow Chemical with a weight-average molecular weight of 300000 g/mol). The concentrations of acrylonitrile in the polystyrenes were:

- 2 (produced by Asahi with a weight-average molecular weight of 204000 g/mol);
- 13.5 (produced by Asahi with a weight-average molecular weight of 149000 g/mol);
- 25 (Tyril 100 produced by Dow Chemical with a weight-average molecular weight of 152000 g/mol);
- 27 (M80 produced by Bayer, no molecular weight information was provided);
- 38 (produced by Monsanto, no molecular weight provided);
- 58 wt.% (produced by Monsanto, no molecular weight was provided).

The equipment and protocol were altered from the work above to accommodate the small sample sizes of some of the copolymers. The small DSM intermeshing, corotating twin-screw extruder that has been described above in earlier work was employed to produce the polymer–montmorillonite composites. The standard condition was 100 r/min at 220°C for 10 min. Higher r/min at 200 and 380°C were also evaluated to determine if shear rate was an independent significant variable for exfoliation. Note that the resident time in the extruder will be constant, irrespective of r/min.

The test samples were prepared by injection molding with a bench-top pneumatic ram molder at 60 psi. The time to file the mold increased with increasing content of acrylonitrile in the polystyrene (0.25 s for pure polystyrene to 0.84 s for 58% acrylonitrile content). The polymer melts were injected at 225°C into an 80°C heated mold. The sample size for testing was $0.155 \times 0.4 \times 3.1$ cm. The montmorillonite content in the composites was 2 and 3.2 wt.%.

WAXS and TEM indicated the expected poor dispersion of the organomontmorillonite in pure polystyrene. The best dispersion of the organomontmorillonite was in the styrene copolymer that contained 38 wt.% acrylonitrile. The tensile modulus of the pure polymers steadily increased from about 3 GPa for polystyrene to about 3.7 GPa for the copolymer with 58 wt.% acrylonitrile. Hence, the slope of the plots of tensile modulus as a function of montmorillonite content is the significant dependent variable that relates to the reinforcement efficiency of the polymers by montmorillonite. The slope increased from about 2.3 for pure styrene to about 6 for the styrene–acrylonitrile copolymer with 58% acrylonitrile content.

The Mori–Tanaka model for an aspect ratio of 100 of perfectly exfoliated montmorillonite aligned in the direction of stress predicts a slope of 10.8 for the same copolymer. When the r/min of the extruder is increased to 200 and 380, the slope increased to about 8.4 for the copolymer–montmorillonite composite with 25 wt.% acrylonitrile content.

The fourth and final article in this series [34] investigates the significance of MA as a copolymer with styrene as regards exfoliation of montmorillonite and mechanical performance. The control was the pure polystyrene, Styron 678 CW manufactured by Dow Chemical. SAN from the previous study was re-evaluated in the range of 2 to 31 wt.% of acrylonitrile content. The SMA copolymers that were evaluated had:

- 2 (Dylark 132 produced by Arco);
- 6 (Dylark 232 produced by Arco);
- 8 (Dylark 232 produced by Arco with a weight-average molecular weight of 240000 g/mol);
- 14 (Dylark 332 produced by Arco with a weight-average molecular weight of 170000 g/mol);
- 25 wt.% (manufactured by Monsanto) content of MA.

Two styrene-acrylonitrile-MA terpolymers; 25% acrylonitrile polymerized with 1% MA and 31% acrylonitrile with 1.3% MA (weight-average molecular weight of 119000 g/mol) were included in the study. Both of these terpolymers were prepared by Bayer.

The same organomontmorillonite that was employed in the previous study was utilized here. The small DSM twin-screw extruder and injection molder that was utilized in the previous study was employed here with the same processing protocol.

WAXS and TEM indicated a similar degree of dispersion for increasing AN and MA content in polystyrene. The pure polystyrene composite indicated a poor dispersion. The higher concentrations of MA in polystyrene seemed to provide an increase in the degree of dispersion when compared to the higher concentrations of AN in the copolymer. The Young's modulus increased steadily from pure polystyrene, approximately 2.96 GPa, to 3.19 GPa for SMA with 14% MA content. Hence, the slopes of modulus as a function of montmorillonite content for each copolymer must be compared to determine the significance of montmorillonite loading as regards an increase in mechanical performance. The slope for the copolymer with 14% MA was approximately 5.2; the slope for the copolymer with 13.5% AN was approximately 3.6.

This increase in reinforcement efficiency appears to be consistent with the WAXS and TEM data. The modulus values for the terpolymers and 90:10 weight ratio blends of SAN:SMA at 25 wt.% were very similar to the modulus value of SAN as 25% AN content.

A review of the work found above with the preparation of polyethylene–montmorillonite composites, polypropylene (polyethylene with pendant methyl groups)–montmorillonite composites, and polystyrene (polyethylene with pendant phenyl groups)–montmorillonite composites indicates that one can balance the hydrophilic–hydrophobic balance of polyethylene and polypropylene with polar functionality with the hydrophilic–hydrophobic balance of organomontmorillonite more effectively than with a similar strategy as regards polystyrene.

In an attempt to mitigate this conundrum, Morgan and Harris [35] employed a solvent (chlorobenzene) blending strategy with organo-montmorillonite exchanged with 1,2-dimethyl-3-*n*-hexadecyl imidazolium ion. Previous work by Gilman *et al.* that describes the preparation of styrene–montmorillonite composites by solvent blending with the same organomontmorillonite [36] and its fire retardant performance is reported in Chapter 8 on flame retardancy.

The montmorillonite that was evaluated was Cloisite Na. The blending strategy included sonication with a Fisher Scientific Sonic Dismembrator Model 500 with a Model 102C Sonic Horn. The organomontmorillonite was first dispersed into the chlorobenzene. The horn of the sonicator was positioned in the montmorillonite dispersion above a magnetic stirring bar. The sonication was at 60% of maximum intensity for 5 min. The polystyrene (Styron 612 manfactured by Dow Chemical) was then dissolved into the montmorillonite dispersion. The polymer–montmorillonite composites (1 and 2.5 wt.% montmorillonite) were prepared by removing the chlorobenzene by rotary evaporation and by subsequent drying under vacuum for 24 h at 80°C.

WAXS and TEM indicate a similar degree of dispersion to the melt-blended SAN–montmorillonite composite with 13.5% acrylonitrile and the SMA–montmorillonite composite with 25% MA discussed in the above work. The montmorillonite is dispersed well in the polystyrene and appears to be intercalated with polymer. Fluorinated synthetic mica (SOMASIF ME-100 manufactured by Co-op Chemical) was treated with the same quat and processes identical to the montmorillonite resulted in significantly inferior results. Sonication was demonstrated to be a significant processing variable for the preparation of improved polystyrene–montmorillonite composites. Unfortunately, no mechanical testing results were provided.

Samakande *et al.* [37] employed a novel quaternary ammonium ion exchanged onto Cloisite Na that had the additional feature of an acrylic functionality. This feature was utilized to polymerize styrene onto the surface of the montmorillonite to form a composite. The quats were (11-acryloyloxyundecyl)dimethyl(2-hydroxyethyl)ammonium bromide (prepared from the reaction of 11-bromoundecylacrylate and dimethylethanolamine) and (11-acryloyloxyundecyl)dimethylethylammonium bromide (prepared from the same bromoacrylate reacting with dimethylethylamine) [38]. High shear is necessary to disperse the montmorillonite from the dry state to individual particles in water. Vigorous stirring (as described in the work) is not sufficient. This is supported by the observation that only 70% of the expected exchange of the quats onto the montmorillonite occurred. This can compromise the exfoliation of the montmorillonite into the styrene monomer.

Bulk polymerization of styrene with the organomontmorillonites occurred with a freshly distilled monomer (all of the inhibitor is removed). Azobisiosobutylnitrile (AIBN) was employed as the initiator. Thermal initiation occurred at 60°C for 72 h.

SAXS and TEM indicated that the montmorillonite exchanged with the quat prepared from dimethylethylamine produced the best polystyrene dispersion. Montmorillonite loading levels were prepared at 2.1, 3.7, 5.7, 7.7, and 18.5 wt.%. Unfortunately, the TEM magnification was not high enough to enable a proper comparison, focused on melt blending and solution blending, with the work above. The distribution of montmorillonite in the 5.7% loaded composite appeared to be very random. The weight-average molecular weight of the polystyrene prepared with the organomontmorillonite at 18.5% loading was 1 120 000 g/mol. The polystyrene was obtained by driving off the polymer from the montmorillonite surface with LiCl. The storage modulus obtained by DMA increased from approximately 8.5×10^8 Pa for pure polystyrene to approximately 1.3×10^9 Pa for the 18.5% loading of montmorillonite in polystyrene.

Further work needs to be done with WAXS, a higher magnification TEM, and tensile modulus evaluations of injected molded test samples before a definite judgment can be made concerning the degree of exfoliation of this organomontmorillonite in a styrene composite prepared by bulk polymerization.

7.4 Mysteries associated with elastomer–clay nanocomposites

Nanocomposites prepared with Cloisite 30B and polyvinylidene fluoride [39] provide an unexpected perspective of the mechanical properties of polymer–montmorillonite nanocomposites. Polyvinylidene fluoride has geminal fluorines on alternate carbons of a polyethylene chain. The polyvinylidene fluoride (PVDF; Kynar 721 manufactured by Atofina Chemical Co.) was premixed as a powder with Cloisite 30B at 5 wt.% with a Flack-Tek DAC-150FV speed mixer. The extrusion of the melt-compounded composite was accomplished with a small DSM twin-screw extruder (capacity 5 cm^3) that was utilized with the SMA and SAN melt-blending work earlier. The temperature of the extruder was 200°C.

Mixing time was not provided. The small injection molder was set with a barrel temperature of 240°C and a mold temperature of 95°C.

WAXS and TEM indicated a good dispersion of highly ordered intercalated montmorillonite that varied in gallery spacing of 2.9 nm. The Young's modulus increased from 1.3 GPa for the pure polymer to 1.8 GPa for the polymer–montmorillonite composite. The percent elongation to failure increased from 20% for the pure polymer to 140% for the polymer–montmorillonite composite. This unprecedented behavior is not explained by the theory developed in Chapter 5. Similar unprecedented behavior will be observed with an unusual polyester prepared with 2,2,4,4-tetramethyl-1,3-cyclobutane diol and a rubber polymer prepared as an acrylonitrile-butadiene (NBR) copolymer.

An extension of this work is found in the second article of this series [40]. The crystal structure of PVDF is altered from primarily α spherulites to β fibers with the preparation of the polymer–montmorillonite composite. Hence, the comparison of mechanical properties between the pure polymer and polymer–montmorillonite composite is skewed by this change of polymer morphology. The T_g of the polymer is low (−40°C). Since the polymer is tested at room temperature, the theory developed in Chapter 5 predicts a greater increase in modulus with montmorillonite loading because of the lower modulus of the pure polymer tested in the rubber region. However, the dramatic increase in percent elongation to failure for the polymer–montmorillonite composite is not adequately addressed by the change in molecular morphology and the lower modulus of the polymer in the rubber region.

The importance of the surface treatment of the montmorillonite in relation to the observations of the phenomenon above is illustrated in Kim and White [41]. Cloisite 20A was used as the organomontmorillonite that was melt blended with PDVF (Solef 1008 manufactured by Solvay). A Brabender internal mixer was utilized to prepare the polymer–montmorillonite composites. The r/min was 100, the mixing temperature was 180°C, and the mixing time was 5 min.

WAXS and TEM were used to characterize the composites. The poor quality of this data in the publication makes interpretation difficult. The composites appear to have intercalated the montmorillonite that was present. The loading levels of Cloisite 20A in the polymer were approximately 3, 5, and 10 wt.%. The increase in Young's modulus as a function of organomontmorillonite was 7% for the 3% montmorillonite loaded composite, 3% for the 5% loaded composite, and 28% for the 10% loaded composite. The percent elongation to failure was reported to decrease. Percent elongation was not reported. Hence, the unexpected mechanical behavior observed above with Cloisite 30B with PDVF was not observed with Cloisite 20A.

Other fluorinated polymers were evaluated in this work:

- vinylidene fluoride-tetrafluoroethylene-perfluoromethyl vinyl ether cure site (P(VDF-TFE-MVE-CSM);
- Viton GLT produced by DuPont Dow Elastomers);
- poly(vinylidene fluoride-hexfluoropropylene) (P(VDF-HFP));
- Viton A 200 produced by DuPont Dow Elastomers);
- poly(vinylidene fluoride-hexafluoropropylene-tetrafluoroethylene) (P(VDF-HFP-TFE));
- Viton B 600 manufactured by DuPont Dow Elastomers);
- poly(tetrafluoroethylene-hexfluoropropylene) (P(E-CTFE), Teflon FEP manufactured by DuPont);
- poly(ethylene-chlorotrifluoroethylene) (P(E-CTFE), Vatar manufactured by Ausimont).

LLDPE 2045 manufactured by Dow Chemical was the control. The significant independent variables for the polymer–montmorillonite composites with the higher mechanical performance were:

- the content of VDF in the copolymer;
- higher dielectric constant;
- high dipole moment.

This observation was expanded into a thermodynamic argument [42] for the prediction of exfoliation of montmorillonite as a function of the surface treatment and the hydrophilic–hydrophobic balance of the polymer continuous phase. More work needs to be done in matching the hydrophilic–hydrophobic balance of fluorinated polymers with the hydrophilic–hydrophobic balance of the organomontmorillonite in order to achieve full exfoliation of the montmorillonite in the polymer. Full exfoliation of the montmorillonite in the polymer will allow for a valid examination of the mechanical properties of fluorinated polymer–montmorillonite composites in relation to the theory of reinforcement developed in Chapter 5.

Vo and Giannelis [43] continued the work above with PVDF and Cloisite 30B. A comparison with Cloisite 20A confirmed that Cloisite 30B is a better choice for improved dispersion efficiency in relation to Cloisite 20A. An examination of the blend of PVDF with nylon 6 with a weight ratio of 30:70 was also provided; PVDF/nylon 6 provided additional insight when compared to the SAN–ABS blend studies above. The blends were prepared with a 3.3% loading of montmorillonite irrespective of the utilization of Cloisite 30B or Cloisite 20A. Nylon 6 (1022B) was supplied by UBE Industries. The PVDF was the same (Kynar 721 supplied by Arkema) as employed in the above studies. The same equipment and protocol were employed in preparing the polymer–montmorillonite composites. The barrel temperature for the microinjector was increased to 275°C and the mold temperature was increased to 110°C.

TEM and WAXS indicated the superior dispersion performance of the Cloisite 30B when compared to Cloisite 20A. The morphology of the polymer blends–montmorillonite composites was significantly different. The particle size of the PVDF dispersed in the nylon 6 without the presence of montmorillonite was approximately 150 nm. When Cloisite 30B was present, the particle size of the dispersed PVDF decreased to approximately 60 nm. When Cloisite 20A was utilized in the blend, the particle size distribution increased significantly to a range of 70 to 150 nm. The impact of the presence of Cloisite 30B on the modulus of the polymer blend was very marked. The modulus of the blend without Cloisite 30B was approximately 1.90 GPa; the modulus of the polymer blend–montmorillonite composite was approximately 2.68 GPa. This value is significantly greater than the PVDF–Cloisite 30B composite (1.36 GPa) and nylon 6–Cloisite 30B composite (2.40 GPa) at the same montmorillonite loading.

The factor for the dramatic improvement in mechanical performance appears to be the successful dispersion and alignment in the direction of stress of the montmorillonite in the dispersed phase (PDVF) and the continuous phase (nylon 6). The percent elongation to failure increased from approximately 55% for the polymer blend without Cloisite 30B to approximately 116% with the presence of Cloisite 30B in the polymer blend. The variability of modulus and percent elongation to failure of the polymer blend with Cloisite 20A was very high and inferior to the polymer blend–Cloisite 30B composite. Blending the Cloisite 30B first with the nylon 6 and then with the PVDF and blending the Cloisite 30B first with PVDF and then nylon 6 produced composites with inferior modulus and percent elongation to break in relation to the blend prepared above by mixing the polymers and Cloisite 30B at the same time. This work provides the promise of superior mechanical performance for elastomer-reinforced polymer blend–montmorillonite composites when the montmorillonite is fully exfoliated and aligned in the direction of stress.

A survey of the modulus as a function of exfoliation efficiency of Cloisite 30B in polyethylene with a varying degree of chlorine content is provided by Kim and White [44]. Polyethylene (2045 LLDPE produced by Dow) was the control. The wt.% chlorine content increased from chlorinated polyethylene (Tyrin; random replacement of hydrogen with chlorine on the polymer backbone; 36% chlorine, manufactured by DuPont Dow Elastomers), polyvinyl chloride (334 FG; 56.7% chlorine, manufactured by OxyVinyl), chlorinated polyvinyl chloride (TempRite, 63.5% chlorine, manufactured by Noveon), to polyvinylidene chloride (Saran, greater than 71.5% chlorine, manufactured by Dow Chemical). The composites were prepared with a Brabender internal mixer at 180°C and 100 r/min. The composites were prepared with 3, 5, and 10 wt.% Cloisite 30B content. The samples for mechanical testing were prepared by compression molding. The composites were characterized by WAXS and TEM.

The quality of the TEM pictures was poor. The dispersions appeared to be intercalated and poorly dispersed. No information was provided with regard to the alignment of the particles in relation to the direction of the applied stress. Because each polymer has a different modulus, the most useful comparison for the determination of reinforcement efficiency of the polymers with Cloisite 30B was the ratio of the modulus of the polymer with and without montmorillonite content. As the content of Cloisite 30B was increased, the polymer with the best modulus response was a function of decreasing chlorine content: chlorinated PVC (3% Cloisite 30B content produced a 15% increase in modulus); PVC (5% Cloisite 30B content produced a 24% increase in modulus); chlorinated PE and PVC has similar moduli at 10% Cloisite 30B content (50% increase). Note that the modulus of the pure polymers decreases significantly as the chlorine content decreases (approximately 2400 MPa for chlorinated PVC; 2000 MPa for PVC; and 4 MPa for chlorinated PE). Percent elongation to failure increases from approximately 900% for the pure polymer to 1000% for chlorinated PE at 5% content of Cloisite 30B.

PVC and chlorinated PVC composites demonstrate a decrease in percent elongation to failure as a function of increasing Cloisite 30B content. A plateau in percent elongation to failure occurs between 3 and 5% Cloisite 30B content for PVC and chlorinated PVC. Additional work is necessary with a twin-screw extruder to prepare improved montmorillonite dispersions and injection molding of the test samples to align the particles in the direction of applied stress in order to investigate the possible synergistic relationship between chlorine content in the polyethylene and an increasing 30B content in the polymer. Although not mentioned in this publication, melt blending of organically treated montmorillonite with PVC results in a significant increase in color (brown). We suspect that this increase in color also occurs with the other chlorinated polyethylenes. This color change limits the utility of organomontmorillonites as a reinforcing dispersed phase for chlorinated polyethylenes, particularly, for the production of clear polymer nanocomposites.

7.5 Dichotomy of crystalline and amorphous polyester–clay nanocomposites

Polyesters include a range of crystalline and amorphous materials similar to that of polyamides. Additional polyester types include unsaturated polymers that can be dissolved into "reactive solvents" such as styrene and methyl methacrylate that participate with the unsaturated polyester in peroxide crosslinking or will crosslink ambiently as coatings that have been dissolved in unreactive solvents.

An amorphous polyester–montmorillonite nanocomposite prepared from condensation of 2,2,4,4-tetramethyl-1,3-cyclobutanediol, 1,3-propanediol, and dimethylterephthalate produced unexpected mechanical properties [45]. Cloisite 20A was melt blended with the polymer at 2.5, 5, and 10 wt.% with a Haake HBI System 90 drive attached to a Rheomex CTW 100 twin-screw extruder. The r/min of the extruder was 60 and the temperature was 120°C. The test samples were prepared in two different injection molders. The test samples from the polyester–montmorillonite composite at 5% Cloisite 20A content were prepared with a Demag Ergotech 35 at 150°C with a pressure range of 10000 to 11000 psi. The test samples from the polyester–montmorillonite composites with 2.5, 5.0, and 10% Cloisite 20A content were prepared with a mini-jector molding machine at a temperature range of 190–200°C.

WAXS indicated an intercalated composite. The modulus values for the 5% loaded composite are indicative of an intercalated composite (a 12% increase above the pure polymer value of 8.66×10^8 Pa). A Halpin–Tsai calculation for a 3% loaded composite predicts a 63% increase in modulus for a fully exfoliated montmorillonite composite with the particles aligned in the direction of applied stress. The unexpected result is the dramatic increase in percent elongation to failure from 68% for the pure polymer to an average of 208% elongation to failure for the composite. The composite samples demonstrate strain hardening and strain whitening past the yield point. The pure polymer does not demonstrate this phenomenon. WAXS indicated no development of strain-induced crystallinity during the test. The WAXS of the composites after strain failure indicated a significant increase in the broadness of the (001)*d*-peak and a shift toward larger *d*-spacing. With the application of heat (a heat gun for 30 s), the test samples return to their original shape and the (001)*d*-spacing pattern that is observed before stress–strain evaluations also returns (shape memory).

Additional information relating to the importance of sample preparation by injection molding is found with the mini-jector prepared samples. Two distinctly different samples were prepared with identical processing conditions for the 2.5% loaded composite. One sample had a distinct region at the surface similar to that obtained with the injection molded nylon 6–montmorillonite composites discussed in chapter 5 above. The effect was the result of higher montmorillonite orientation in the molding direction because of the high shear environment near the surface of the mold. The samples are smaller than the standard size "dog bones" prepared with the Demag Ergotech (see the description of the DSM mini-extruder and injection molder found above). The thickness of the outer region is about a quarter of the sample thickness on both sides of the sample.

The other sample type did not have this region. The sample with the separate region had a modulus of approximately 12.38×10^8 Pa and a percent elongation to failure similar to the pure polymer (82.8% for the composite and 68% for the pure polymer). An example of the sample without the distinct region at the surface had a modulus of 11.56×10^8 Pa and a percent elongation to failure of 365.3%.

The WAXS of the sample with high elongation was virtually featureless; the sample with the higher modulus and lower elongation had the expected diffraction pattern of an intercalated composite. Compression molding both samples to thin films produced composites that had identical WAXS and the expected pattern of an intercalated composite. The outer region of the sample with higher modulus and less percent elongation to failure has a significantly different montmorillonite orientation and organization than the sample with the lower modulus and higher percent elongation to failure.

The differences in the injection molding processing seem to be a significant variable in controlling the construction of the polymer composite. The modulus of the 5% loaded composite samples prepared with the smaller injection molder increased to an average of 13.6×10^8 Pa with an average percent elongation to failure of 120 ± 59%. This is consistent with an improved dispersion and orientation of the intercalated montmorillonite particles in the direction of applied stress. The 10% loaded composite provided a mechanical performance with a higher modulus (24.8×10^8 Pa) and significantly lower percent elongation to failure (17 ± 17%) anticipated with an intercalated polymer–montmorillonite composite.

Further work is necessary to deconvolute the role of the apparent significant independent processing variables and particle morphology. Solid state nuclear magnetic resonance (NMR) should be valuable in detecting and measuring the degree of the development of apparently different polymer morphologies that result as a function of these processing variables.

Polyethyleneterephthalate (PET) is an example of a crystalline polyester with commercial significance. Work by Kracalik *et al.* [46] evaluated the importance of melt-blending organomontmorillonites with recycled PET in adding value to the potentially important source of the polymer. The organomontmorillonites (Cloisite 25A, 10A, and 30B) were evaluated in the polymer at 5 wt.% based on montmorillonite content.

Silane ([3-(glycidyloxy)propyl]trimethoxysilane, hexadecyltrimethoxysilane, and (3-aminopropyl)trimethoxysilane) modification for improved dispersion efficiency of the organomontmorillonites was also evaluated. The silane modifications were prepared by dispersing the organomontmorillonites in a methanol–water (10:1) mixture, adding 0.3g of silane to the dispersion, and subsequent agitation for two days at room temperature. The silane-modified organomontmorillonite was separated from the solution by filtration, repeatedly washed with methanol, and dried at 50°C for 4h.

Care should be exercised with this procedure. Quaternary ammonium ion can be extracted from the montmorillonite with methanol (particularly the Cloisite 10A, which is exchanged past the CEC). The organic content needs to be checked by thermal analysis after this procedure. The recycled PET was a color-sorted grade from Polymer Institute Brno. The DSC signature of the PET was confirmed as similar to virgin PET. The mini-extruder and injection molder from DSM were used to prepare the polymer–montmorillonite composites and test samples:

1. The extruder was set at 255°C.
2. The composites were prepared at 200 r/min;

The residence time in the extruder was 10 min.

3. The injection molder was set at 260–265°C.
4. The injection molder dwell time was 10 s.

Unfortunately, the samples for TEM evaluation were prepared by compression molding (260°C for 5 min with a Collin 200P press). There is no assurance that the particle alignment of the test samples will be similar to the compression-molded samples.

WAXS and TEM indicated intercalated composites. The composite with the highest modulus was prepared with Cloisite 25A (2984 Mpa versus 2170 MPa for the pure polymer). As expected, the percent elongation to failure decreased from 316.5% for the pure polymer to 30.6% for the composite.

Montmorillonite has proven to be a very efficient nucleating agent for PET. An evaluation of the role of the increase of crystallinity for the composites on mechanical properties needs to be carried out to complete the study.

Another class of crystalline polyesters with commercial potential is biodegradable polyesters. Pollet *et al.* [47] evaluated three different biodegradable polyesters (polybutylene succinate (BIO; Bionolle manufactured by Showa High Polymer Co.), and two polybutylene-adipate-co-butylene-terephthalates (ECO; Ecoflex F manufactured by BASF and EAS; EastasrBio Ultra manufactured by Novamont)). The organo-montmorillonite utilized in the evaluations was Cloisite 30B. The Cloisite 30B was dried at 40°C under reduced pressure for 4 h. Cloisite 30B must be properly dried for this application. The composites with the best mechanical performance were prepared with a two-roll mill (Agila) for 10 min with the temperature set at 150°C for EAS and ECO and 160°C for BIO. The montmorillonite content of the composites was 3 wt.%. Preparing a master batch of polymer with a 36 wt.% Cloisite 30B initially with subsequent formulation was not as effective.

Unfortunately, the test samples were prepared by compression molding. The pressing protocol was 170°C at a pressure of 150 bar for 3 min, 30 bar for 3 min, and finally, 30 bar at 15°C for 5 min. The best-performing polymer–montmorillonite composite was prepared with BIO. The modulus increased from 323 ± 16 MPa for the pure polymer to 405 ± 19 MPa for the composite. The percent elongation to failure decreased from 220 ± 8 percent for the pure polymer to 149 ± 10 percent for the composite.

Because the quat exchanged onto the montmorillonite has two primary hydroxyl groups per quat, the impact of transesterification of the polyester with the hydroxyl groups on the modulus was evaluated. Three transesterification catalysts were evaluated (dibutyl tin dilaurate, titanium (IV) butoxide, and antimony (III) oxide) by formulating them directly into the formulation prepared with the two-roll mill at 1 wt.%. Only the tin catalyst produced a polymer–montmorillonite composite with a significant increase in modulus (515 ± 11 MPa). The percent elongation to failure decreased to 66 ± 19 percent. Additional work needs to be done to optimize the amount of hydroxyl functionality on the montmorillonite (too much functionality will lower the molecular weight of the polyester and compromise performance). All of the components of the formulation must be carefully dried for the most effective transesterification. The test samples must be prepared by injection molding.

Unsaturated polyesters are designed to provide rigid structures. There is not enough flexibility in commercial unsaturated polyesters to compensate for the compromise in flexibility when montmorillonite is employed as a nanoparticle reinforcement. Hence, the full realization of the value of montmorillonite as a nanoparticle reinforcement in unsaturated polyesters has not been achieved.

The dispersion of montmorillonite in unsaturated polyesters produces a very viscous fluid. Alignment of the montmorillonite in the direction of applied stress in the cured composite is problematical. References [48–52] represent a large selection of publications representing this conundrum. This issue is reminiscent of the viscosity issues discussed in Chapter 6 with epoxy–montmorillonite composites.

With the advent of NanoXcel technology for the Yamaha FX HO WaveRunner personal watercraft [53], montmorillonite as a reinforcing dispersed phase in unsaturated polyesters has become a reality. The new technology is based on the utilization of organomontmorillonite Cloisite 30B to form a polymer nanocomposite with a hybrid unsaturated polyester-urethane polymer supplied by Reichhold Inc. This is a marine-grade unsaturated polyester.

The new polymer nanotechnology allowed for the removal of calcium carbonate from the formulation which lowered the conventional polymer composite density from 1.9 g/cm^3 to 1.45 g/cm^3. This resulted in a 25% reduction in the weight of the watercraft.

Interplastics Corporation Molding Products Division (formulator of the polymer nanocomposite sheet-molding compound (SMC) for Yamaha) and Yamaha Watercraft Group received the 2008 JEC Innovation Award in the Sports and Leisure category for this novel technology. In addition to the weight savings, the technology provides a low-pressure, Class A surface finish and high-impact resistance.

Urethane modification of unsaturated polyesters for the effective utilization of montmorillonite as a reinforcing nanoparticle for unsaturated polyesters is not a panacea. You *et al.* [54] prepared a hydroxylethyl methacrylate terminated polyurethane from 2,4(2,6)-toluene diisocyanate and polyethylene glycol ($M_n = 2000$) isocyanate prepolymer. The grade and manufacturer of the unsaturated polyester were not specified. The organomontmorillonite was prepared from the exchange of trimethyl hexadecylquaternary ammonium bromide onto an aqueous dispersion of the montmorillonite. The exchange capacity of the montmorillonite and the amount of quat exchanged onto the montmorillonite were not documented. The isocyanate prepolymer, styrene, and the organomontmorillonite were dispersed together. The unsaturated polyester was added to the dispersion. Finally, the hydroylethyl methacrylate terminated polyurethane unsaturated polyester was cured with methyl ethyl ketone (MEK) peroxide and cobalt naphthoate catalyst (supplied by Guangzhou Nanfei Trading Co.). The amount of catalyst and MEK peroxide was not specified. The test samples were prepared by curing in a mold at room temperature for 24 h; final cure was at 80 °C for 4 h.

The polymer–montmorillonite composites were characterized by TEM and WAXS. The montmorillonite was intercalated and poorly dispersed in the polymer. The amount of acrylic-urethane content in the unsaturated polyester varied from 0 to 30%. Acylic-urethane content was a statistically significant variable for decreasing the tensile strength of the composite. The maximum reinforcement was achieved at 3 wt.% montmorillonite content; additional montmorillonite content up to 5 wt.% compromised mechanical performance. The tensile strength was generally a 50% increase at 3% loading versus the polymer without montmorillonite for all acrylic-urethane contents. The best-performing composite was without acrylic-urethane modification with an increase in tensile strength from approximately 45 MPa for the pure polymer to approximately 62 MPa for the polymer–montmorillonite composite with 3% loading.

7.6 Two-phase engineered polymer (polyurethane) synergy with clay nanocomposite reinforcement

Polyurethanes are noted for their unique mechanical toughness. The combination of “hard” and “soft” segments in the polymer chain can lead to segregation of the hard segments into separate domains dispersed in the soft segment continuous phase. The size of the dispersed polymer phase is a significant variable as regards the unique toughness of polyurethane. Zhang *et al.* [55] prepared polyurethanes with Cloisite 10A by first dehydrating the organomontmorillonite with boiling toluene at 110°C for 6 h. This is an effective method for removing water from montmorillonite. Even the organomontmorillonites are hydroscopic. Water will react rapidly with isocyanate and compromise the production of polyurethane. One must be cautious with this technique. Toluene at any temperature can remove quat from the surface of montmorillonite, particularly, organomontmorillonites that have quat treatments in excess of the exchange capacity of the montmorillonite.

Cloisite 10A has an excess of quat of 30 meq/100 g of montmorillonite past the exchange capacity. The organic content of the organomontmorillonite needs to be determined after the dehydration. This apparently was not done. Dried Cloisite 10A was dispersed into polyethylene glycol (0.05 mol). Toluene diisocyanate (0.15 mol; distilled before use) was added to the dispersion with stirring. The reactor was purged with nitrogen (presumably dry). The reaction was increased in temperature to 80°C for 4h. This isocyanate prepolymer was chain extended by the slow addition of 1,4-butanediol. Final conversion was accomplished at 90°C under vacuum for 15 h. Dog-bone samples ($10 \times 2 \times 2$ mm) were produced for the mechanical testings. The process for their preparation was not described.

WAXS and TEM were employed to characterize the polymer–montmorillonite composite. The composites were intercalated and appeared to have a poor dispersion. The tensile strength of the polyurethane increased from approximately 0.6 MPa without montmorillonite to approximately 1.5 MPa with a loading of approximately 7 wt.% montmorillonite. The percent elongation to failure of the polyurethane increased from approximately 60% without the presence of montmorillonite to approximately 170% with a loading level of approximately 7% montmorillonite in the polymer. This contiguous increase of tensile strength and percent elongation at failure as a function of increasing montmorillonite content in polyurethane appears to be systemic with polyurethane–montmorillonite composites and is not duplicated with other dispersed-phase reinforcements.

Instead of the diols above, Maji *et al.* [56] evaluated a hydroxyl functional polyester–polyurethane prepolymer prepared by the ester formation of diethylene glycol with adipic acid and subsequent urethane formation with the 2,4-toluene diisocyanate (TDI) reaction with the excess hydroxyls on the polyester (Urepan 600 prepared by M/s Rhein

Chemie Rhenau GmbH). The excess hydroxyl content of the prepolymer was determined at 25.8 mg KOH/g of polymer based on Karl Fischer titration. The prepolymer was dissolved in THF (10 g of polymer per 90 g of THF) by soaking the polymer in the solvent overnight and then by high-speed dispersion for 1 h. This was accomplished at room temperature.

Cloisite 30B was employed as the organomontmorillonite of choice for this investigation. Cloisite 30B has the feature of two primary hydroxyl groups per quat that is exchanged onto the montmorillonite. These primary hydroxyl groups have the potential to cure with the TDI to form polyurethane functionalities. Cloisite 30B was dispersed in THF at a temperature range of 0 to 10°C with a sonicator for 30 min. The organomontmorillonite–THF dispersion was dispersed into the prepolymer–THF dispersion. TDI was added to the organomontmorillonite–prepolymer THF dispersion with the functionality exactly matching the hydroxyl content of the prepolymer plus an excess to crosslink the polymer through the formation of allophanates with the nitrogen of the urethane linkage. The Cloisite 30B concentrations in the final urethane polymer were evaluated at 1, 3, and 5 wt.%. The test samples were die cut from cast sheets. The cast sheets were prepared by THF evaporation from the polyurethane–montmorillonite dispersions at room temperature. WAXS and TEM were employed to characterize the polyurethane–montmorillonite dispersions.

The 1 and 3% loaded samples appear to have intercalated and exfoliated composites with good dispersion. The 5% loaded sample was intercalated with a poor dispersion. The modulus of the composites was evaluated as a function of percent elongation during the stress–strain evaluation. At 100% elongation, the modulus of the 3% montmorillonite loaded sample increased from 10.80 ± 0.23 MPa for the pure polyurethane to 16.68 ± 021 MPa. The tensile strength increased from 15.57 ± 0.36 MPa for the pure polymer to 30.85 MPa for the composite. The percent elongation to failure increased from 502 ± 10% for the pure polymer to 581 ± 13% for the composite. The 5% loaded composite had an increased modulus of 20.83 ± 0.22 MPa; a decreased tensile strength below the 3% composite (28.74 ± 0.40 MPa); and a percent elongation to failure below the pure polymer (438 ± 17%). One might imagine that an increased loading of montmorillonite with good exfoliation and dispersion in excess of 3% loading in this polymer would achieve remarkable toughness (increasing modulus, tensile strength, and percent elongation) not duplicated by any other dispersed-phase reinforcement.

In an attempt to achieve this high level of toughness, Kaushik *et al.* [57] prepared a layer-by-layer construction of montmorillonite and a cationic polyurethane. A construction of 300 layers was prepared and tested. The cationic polyurethane was prepared from isophorone diisocyanate, poly (tetraethylene glycol), and 3-diethylamino-1,2-propanediol. The isocyanate prepolymer was chain extended with ethylene diamine. The cationic polyurethane was supplied by HEPCE CHEM Co. Ltd. as a 35 wt.% aqueous dispersion. The molecular weight was reported to be 90000 (no reference was made to the weight being number or weight average).

An aqueous dispersion of Cloisite Na^+ was prepared at 0.5 wt.% by "vigorous stirring" of the montmorillonite in deionized water (18 MΩ cm^{-1}, pH = 5.6). High shear is required to disperse Cloisite Na^+ properly into water. The dispersion was allowed to set for one week. The supernatant was separated from the settled montmorillonite and utilized to prepare the layer-by-layer construction. Cleaned glass slides were used to stage the construction. The glass slide was first emersed in the polymer emulsion for 5 min. The slide was removed, rinsed with deionized water for 2 min, and dried with a compressed air stream for 1 min. The slide was then immersed into the montmorillonite dispersion for 5 min, rinsed with deionized water for 2 min, and dried with a compressed air stream for 1 min. This was considered to be one layer. The cationic polyurethane will exchange onto the montmorillonite in a similar fashion to quat exchange to prepare organomontmorillonites. This process was repeated on 300 prepared layers. The glass was etched away from the polymer–montmorillonite construction with HF. The films were soaked in isopropanol overnight and then dried at approximately 80 °C. The montmorillonite content of the construction was varied by dilution of the polymer dispersion before each construction was prepared. The wt.% content of the different constructions was 13, 18, 21, 27, and 41%.

SEM indicated well-defined regions of segregated layers of montmorillonite and polymer. The bilayer thickness decreased respectively with montmorillonite content by 53, 31, 24, 17, and 11 nm. The modulus of the constructions increased from 0.025 ± 0.005 GPa for the pure polymer to 3.6 ± GPa for the construction with 41 wt.% montmorillonite. The tensile strength increased from 13.2 ± 0.9 MPa for the pure polymer and plateaued at approximately 39.4 ± 3.0 to 41.2 ± 3.8 MPa in the range of 13 to 21 wt.% montmorillonite content in the construction. The percent elongation to failure steadily decreased from 4.10 ± 0.10 for the pure polymer to 0.008 ± 0.003 for the construction with 41 wt.% montmorillonite. Apparently, the restriction of the polymer in the narrow galleries of the construction, coupled with the tethering of the cationic sites on the polymer to the surface of the montmorillonite, significantly compromised the toughness of the polymer.

In order to explore the role of domain sizes in polyurethane–montmorillonite composites, Maji *et al.* [58] introduced hyperbranched polyols (HBPs) of different sizes into the urethane synthesis. These are polyesters with terminal, primary hydroxyl groups at the edges of the HBP. They increased in size from H20 ($M_w = 2100$ g/mol; polydispersity = 1.3), to H30 ($M_w = 3500$ g/mol; polydispersity = 1.5), and the largest being H40 ($M_w = 7316$ g/mol; polydispersity = 1.8). The HBPs were obtained from Perstorp Specialty Chemicals AB. The novel polyurethanes were prepared by reacting polypropylene glycol (molecular weight (number-average or weight-average was not designated in the work) of 2000 from Sigma-Aldrich Chemicals) with TDI (toluene 2,4-diisocyanate from Merck Schuchardt OHG) with a 1 mol excess of TDI under a nitrogen atmosphere (presumably dry) at 80°C for 4 h. The isocyanate prepolymer has a 4.4% excess of isocyanate. Each HBP was reacted with the isocyanate prepolymer separately. The isocyanate functionality to hydroxyl functionality of the HBP was maintained at a 1:1 ratio. The solvent was dry THF (from Rankem); the reaction was run at room temperature; and 0.001 moles of dibutyltin dilaurate catalyst was employed. Cloisite 30B was dispersed in THF with sonication and added to the prepared polyurethane. Note that, presumably, the primary hydroxyl groups on the quat exchanged onto the montmorillonite will not have an opportunity to cure with the isocyanate. The concentrations of Cloisite 30B in the polymer were 2, 4, 8, and 16 wt.%.

WAXS and TEM were utilized to characterize the composite. The montmorillonite was intercalated and exfoliated; the 16 wt.% composites appear to have more intercalated structures. The dispersions appear to be good. The test samples were die cut from a cast sheet. The sheet was prepared by evaporation of THF from the polymer solution at room temperature on a circular quartz mold. Each polyurethane without montmorillonite had a different modulus at 100% elongation (with H20, 2.84 ± 0.15 MPa; H30, 2.65 ± 0.12 MPa, H40, 3.56 ± 0.11 MPa).

Hence, a ratio of the modulus with montmorillonite present divided by the modulus without montmorillonite is the most useful comparison. The performance of the polyurethanes with 16 wt.% montmorillonite indicates a compromise in flexibility illustrated by a decrease in percent elongation to failure compared to the 8 wt.% polyurethane–montmorillonite composites. The polyurethanes with up to 8 wt.% montmorillonite appear not to compromise flexibility (this is consistent with the WAXS and TEM measurements). Unfortunately, the ratios of the polyurethane composites prepared with H30 were not provided (the ratios with H20 and H40 are illustrated in Figure 9 on page 298 of reference [58]).

The best-performing composite was prepared with H30 at 8 wt.% montmorillonite. The ratio 1.83 indicated the largest enhancement of modulus relative to the pure polyurethane for the 8 wt.% containing composites. The percent elongation to failure increased from 236 ± 15% for the pure polymer to 374 ± 32% for the 8% loaded polyurethane–montmorillonite composite. These hyperbranched polyols provide a significant enhancement of mechanical properties for the polyurethane cure chemistry evaluated with Cloisite 30B as the dispersed-phase reinforcement. Additional work needs to be done with Cloisite 30B present during the cure to ascertain if further benefits to mechanical properties are derived from the reaction of the isocyanate with the primary hydroxyls on the quat exchanged onto the montmorillonite.

7.7 Elastomers that crosslink with clay nanocomposite reinforcement

Mechanical evaluations of rubber–montmorillonite composites provide a common theme. Full exfoliation of montmorillonite in rubber is difficult. Enhancement of mechanical properties is greater than with thermoplastic polymers, because the difference in the modulus of the continuous rubber phase in relation to the montmorillonite dispersed phase is greater. Percent elongation to failure increases, modulus increases, and tensile strength increases as the concentration of montmorillonite increases in the rubber.

An excellent example of this is found in the patent application by Powell [59]. A Ferro Lab Banbury mixer was utilized to prepare a nitrile rubber (Nipol 150; 41% nitrile content) montmorillonite composite. The mixing temperature was maintained under 120°C; the fill factor was 60%. Cloisite 10A was added to the mixing rubber in three increments over a 15 min period. Zinc stearate was added phr after all of the organomontmorillonite was added. The combination of zinc oxide and stearic acid was not used, so as to eliminate the addition of a larger sized dispersed phase (zinc oxide) that could compromise mechanical performance. The zinc stearate–montmorillonite–nitrile rubber dispersion was removed from the mixer and passed through a two-roll mill six times to finish the dispersion. The dispersion was returned to the Banbury. Sulfur

(3.5 phr) and 2,2-benzothiazole disulfide (2 phr) were added to the dispersion in the Banbury. The temperature was maintained at below 90 °C with a mixing time of 3 min. This final formulation was removed from the Banbury and finished by passing through a two-roll mill six times. The rubber composite was cured at 150 °C in a press for 30 min at a pressure of 2400 psi on a 16 in square plate. The Cloisite 10A content was evaluated at 5, 10, 15, 20, 25, 30, 40, and 50 phr. Carbon black (NS 774) was the control dispersed phase with the same nitrile rubber. The processing parameters were identical to those used above for Cloisite 10A, except that the standard zinc oxide (2 phr; Kadox 920 manufactured by Zinc Corporation of America) and stearic acid (2 phr) were utilized in the formula. The carbon black levels in the formula were evaluated at 15, 25, 55, 80, and 100 phr. The test samples were die cut from the cured rubber slabs.

As expected, the modulus for the carbon black control could only be measured at 100% elongation past a loading level of 25 phr. The modulus at 100% elongation for the carbon black controls increased as a function of carbon black loading to approximately 2589 psi; the elongation to failure was approximately 138%. The modulus for the Cloisite 10A composites increased to approximately 1000 psi for the 50 phr loading and (in apparent defiance of the classic physics model for dispersed-phase reinforcement) the percent elongation to failure increased with Cloisite 10A loading to 368% for the 50 phr loaded rubber composite. The modulus at 300% elongation of the Cloisite 10A composite at 50 phr loading was approximately 2200 psi. The tensile strengths for the carbon black composites plateaued at approximately 80 phr loading with a value of approximately 3100 psi; the tensile strengths of the Cloisite 10A composites continued to increase as a function of montmorillonite loading to a similar maximum value for 50 phr montmorillonite loading of approximately 3175 psi.

Hence, the highest loading of montmorillonite in the rubber composite was significantly tougher than the carbon black rubber composites starting at a loading level of 55 phr up to 100 phr. The maximum value of modulus, tensile strength, and percent elongation of the Cloisite 10A rubber composite was never determined.

Characterization of the montmorillonite–rubber composites was carried out using WAXS, TEM, and SEM. The TEM and SEM results were not reported in the patent application. The montmorillonite was intercalated and highly aligned in the direction of applied stress. The WAXS peaks were narrow and indicated a (001), (002), and (003)*d*-diffraction pattern.

In order to circumvent the problematical nature of preparing fully exfoliated montmorillonite in rubber by compounding, Sadhu and Bhowmick [60] used rubber solutions for the dispersion of the montmorillonite. NBR (19 and 50% acrylonitrile content supplied by Bayer; 34% acrylonitrile content supplied by Apar), styrene-butadiene rubber (SBR; 23% styrene; Synaprene 1502 supplied by Synthetics and Chemicals, Ltd), butadiene rubber (BR; the supplier was not identified) were dissolved in solvents (toluene for SBR and BR; chloroform for NBR). Organomontmorillonite prepared by exchanging an aqueous dispersion of Cloisite Na^+ (high shear is necessary to prepare an optimal aqueous dispersion; shear conditions were not provided) with the ammonium ions of octadecyl amine (acidified with HCl). The organomontmorillonite was dispersed in ethanol. The organomontmorillonite–ethanol dispersion was added to the rubber solutions and stirred for 2 h. Dicumyl peroxide was added to the rubber–montmorillonite dispersion at 1 phr. The solvents were removed under vacuum at 50°C. The resulting composite was dried under these conditions for two days. The resulting montmorillonite–rubber composites were passed through a two-roll mill to prepare 1 mm sheets for testing and cured at 160°C. Each rubber composite was prepared with 2, 4, 6, and 8 wt.% organomontmorillonite.

The test samples were die cut from these sheets. Characterization was carried out by WAXS and TEM. The quality of the TEM pictures is poor. The montmorillonite appears to be intercalated in the rubber composites. The best-performing NBR composite appears to be with the rubber that contains 34% acrylonitrile. The Young's modulus increased from 11.1×10^3 kPa for the pure polymer to 12.5×10^3 kPa for the 4% loaded composite. The percent elongation to failure increased from 399% for the pure polymer to 923% for the 4% loaded composite. Tensile strength increased from 2.10 MPa for the pure polymer to 4.85 MPa for the 4% loaded composite. This NBR–montmorillonite composite is significantly tougher than the pure polymer. The work done at break increased from 2138 J/m^2 for the pure polymer to 8268 J/m^2 for the 4% loaded composite.

When the rubber is prepared by emulsion polymerization, e.g. SBR, or harvested as an aqueous emulsion (natural rubber; NR), an effective process for preparing excellent rubber–montmorillonite composites consists of blending an aqueous dispersion of the montmorillonite with the aqueous polymer dispersion [61]. This dispersion is stable, because the stabilization mechanism (anionic stabilization) is identical for the latex and montmorillonite. This dispersion can be processed with the same commercial process protocol employed to produce the rubber crumb (colloid destabilization with acid and or salts), which can then be bailed for rubber compounding. It is critical that the montmorillonite be optimally dispersed in the water. The work in the cited reference above utilized the montmorillonite aqueous dispersion that was sampled from the commercial process for the production of Cloisite products. The aqueous dispersion was produced with intense shear conditions and purification protocol. The dispersion was characterized for optimal individual particle separation by low-angle laser light-scattering measurements.

A variation on this theme involves exchanging a reactive quat, N-allyl-N,N-dimethyl-octadecyl ammonium chloride (from the Applied Chemistry Laboratory at Beijing University of Chemical Technology) onto the montmorillonite (CEC = 78 meq/100g of clay from Liufangzi Clay Factory) as a dispersed phase in water (2 wt.% montmorillonite content).

The dispersed-phase stability of the organomontmorillonite [62] was not compromised. The weight percent of quat on the montmorillonite was evaluated at 0.1, 0.15, 0.2, and 0.3%. This quat will participate with the cure chemistry during vulcanization. The aqueous SBR latex (22.4% solids supplied by Jilin Petrochemical Co. Ltd.) was blended with the organomontmorillonite aqueous dispersion. The polymer–functional montmorillonite dispersion was flocculated with 2% sulfuric acid, washed with water, and dried in an oven at 50°C for 20h. The rubber crumb was formulated at 10 phr montmorillonite content. The rubber was compounded with a two-roll mill and cured at 150°C. The compound ingredients appear to be standard with standard concentrations.

The SBR–functional montmorillonite composites were characterized by WAXS and TEM. The composites before vulcanization appear to be exfoliated and intercalated with a good dispersion of particles. The vulcanized composite appears to present a compromised exfoliation, intercalation, and dispersion. The ingredients in the formula appear to be responsible for this change. The modulus at 100% elongation for the composite with montmorillonite and no reactive quat increased from 1.4 MPa to 1.6 MPa with a 0.3% content of the reactive quat on the montmorillonite. The percent elongation to failure increased from 592% without reactive quat to a maximum of 734% elongation to failure for the montmorillonite with 0.1% reactive quat exchanged on the montmorillonite. Additional reactive quat on the montmorillonite compromised performance. Tensile strength achieved a maximum value of 18.7 MPa (compared to 4.3 MPa for the montmorillonite composite without the quat) with an exchange of 0.2% reactive quat.

Relatively small amounts of the reactive quat exchanged onto the montmorillonite provided a dramatic increase in rubber toughness when compared to the montmorillonite–rubber composite without the quat. There are many more examples in the literature (patents and articles) that echo the above theme. When montmorillonite is properly intercalated or exfoliated into rubber, the modulus, percent elongation to failure, and tensile strength increase significantly as a function of increasing montmorillonite concentration in the rubber. The result is a dramatic increase in the toughness of the rubber–montmorillonite composite.

7.8 Conclusions

The preparation of polyolefin–montmorillonite nanocomposites presents a complex perspective as regards the successful exfoliation and alignment of montmorillonite in the direction of applied stress in the final construction. The variability of polyethylene and polypropylene from commercial suppliers is a significant independent variable, particularly as regards successful melt blending to prepare polymer nanocomposites. Hence, there is no panacea for the process.

Twin-screw extruders with excellent mixing, shear, and residence time are critical for successful preparations. Injection molding at high pressures through the gate of the mold is critical. Cloisite 20A is the closest organomontmorillonite to a universal solution for this polymer type. Polar modification of the polyolefin is beneficial to the successful exfoliation of montmorillonite. These modifications add cost and compromise polymer performance in relation to the decreasing molecular weight of the polyolefin.

TPO adds an additional complexity to the process of preparing polymer nanocomposites. The apparent surface energy associated with the interface between the dispersed rubber phase (polyethylene) in polypropylene will provide segregation regions for the organomontmorillonite. This will prejudice the preferred alignment of the montmorillonite and compromise performance.

Because of the high temperatures associated with the preparation of polyimides, traditional organomontmorillonite technology associated with commercial products is not suitable for utilization in the preparation of polyimide–montmorillonite nanocomposites. The thermal stability of the quaternary ammonium ions present at the interface of the montmorillonite is not sufficient for successful implementation with the polyimide synthesis. Clever adoption of higher temperature stable surface treatments will result in successful montmorillonite–polymer nanocomposites.

The successful preparation of polystyrene–montmorillonite nanocomposites is very troublesome. More needs to be known about the structure–property relationships that exist with polystyrene in conjuction with the introduction of montmorillonite nanoparticles into the polymer before proper nanocomposites can be prepared.

Elastomers (rubber) have the advantage of lower T_g than polymers for engineered applications and equipment uniquely designed for the preparation of rubber compounds that allows for the incorporation of much larger amounts of montmorillonite without a detrimental sacrifice of the hardness–flexibility balance of the nanocomposites. The alignment of intercalated composites for the maximum benefit of mechanical properties is much easier.

Polyesters provide the option of nanocomposites prepared with crystalline and amorphous continuous phases. The crystalline polymers behave in a similar way to the polyamides (nylons) as regards structure–property relationships. The amorphous polymers are a different matter. What is unexplained by first-principle arguments is the dramatic increase in percent elongation to failure [39,45,59] observed with amorphous polymer–montmorillonite composites. This phenomenon also applies to elastomers.

The advent of polyurethane–montmorillonite nanocomposites presents the unique opportunity to integrate two phases (soft continuous phase with a harder dispersed phase) with a montmorillonite-dispersed nanoparticle. The opportunity to provide functionality that will cure with the urethane on the surface of the montmorillonite results in a unique composite with mechanical properties that is not duplicated by other polymer–montmorillonite nanocomposites.

An additional advantage to those described above with elastomeric (rubber)–montmorillonite nanocomposites is the ability to crosslink the elastomer with sulfur or peroxide. A dramatic increase in toughness (improved elasticity) as the result of the proper management of crosslink density and montmorillonite loading in these curable polymers is realized.

References

[1] S. Hotta and D. R. Paul. Nanocomposites formed from linear low density polyethylene and organoclays. *Polymer*, **45** (2004), 7639–7654.

[2] R. K. Shah and D. R. Paul. Organoclay degradation in melt processed polyethylene nanocomposites. *Polymer*, **47** (2006), 4075–4084.

[3] R. K. Shah, D. L. Hunter, and D. R. Paul. Nanocomposites from poly(ethylene-co-methacrylic acid) ionomers: effect of surfactant structure on morphology and properties. *Polymer*, **46** (2005), 2646–2662.

[4] T. D. Fornes, P. J. Yoon, D. L. Hunter, H. Keskkula, and D. R. Paul. Effect of organoclay structure on nylon 6 nanocomposite morphology and properties. *Polymer*, **43** (2002), 5915–5933.

[5] R. K. Shah and D. R. Paul. Comparison of nanocomposites prepared from sodium, zinc, and lithium ionomers of ethylene/methacrylic acid copolymers. *Macromolecules*, **39** (2006), 3327–3336.

[6] R. K. Shah, D. H. Kim, and D. R. Paul. Morphology and properties of nanocomposites formed from ethylene/methacrylic acid copolymers and organoclays. *Polymer*, **48** (2007), 1047–1057.

[7] L. Cui, C. Troeltzsch, P. J. Yoon, and D. R. Paul. Morphology and properties of nancomposites formed from poly(ethylene-co-methacrylic acid) ionomers and organoclays: effect of acid neutralization. *Macromolecules*, **42** (2009), 2599–2608.

[8] L. Cui, X. Ma, and D. R. Paul. Morphology and properties of nanocomposites formed from ethylene–vinyl acetate copolymers and organoclays. *Polymer*, **48** (2007), 6325–6339.

[9] I. S. Suh, S. H. Ryu, J. H. Bae, and Y. W. Chang. Effects of compatibilizer on the layered silicate/ethylene vinyl acetate nanocomposite. *Journal of Applied Polymer Science*, **94** (2004), 1057–1061.

[10] M. Tillekeratne, M. Jollands, F. Cser, and S. N. Bhattacharya. Role of mixing parameters in the preparation of poly(ethylene vinyl acetate) nanocomposites by melt blending. *Journal of Applied Polymer Science*, **100** (2006), 2652–2658.

[11] S. K. Swain and A. I. Isayev. Effect of ultrasound on HDPE/clay nanocomposites: rheology, structure and properties. *Polymer*, **48** (2007), 281–289.

[12] R. K. Shah, L. Cui, K. L. Williams, B. Bauman, and D. R. Paul. Nanocomposites from fluoro-oxygenated polyethylene: a novel route to organoclay exfoliation. *Journal of Applied Polymer Science*, **102** (2006), 2980–2989.

[13] C. Guo, Z. Ma, M. Zhang, A. He, Y. Ke, and Y. Hu. Preparation of PE/MMT nanocomposite by monomer intercalation and *in situ* copolymerization. *Chinese Science Bulletin*, **47**:15 (2002), 1267–1270.

[14] J.-H. Lee, D. Jung, C.-E. Hong, K. Y. Rhee, and S. G. Advani, Properties of polyethylene-layered lilicate nanocomposites prepared by melt intercalation with a PP–g–MA compatibilizer. *Compositer Science and Technology*, **65** (2005), 1996–2002.

[15] J. H. Kim, C. M. Koo, Y. S. Choi, K. H. Wang, and I. J. Chung, Preparation and characterization of polypropylene/layered silicate nanocomposites using an antioxidant. *Polymer*, **45** (2004), 7719–7727.

[16] Y.-Q. Zhang, J.-H. Lee, J. M. Rhee, and K. Y. Rhee. Polypropylene–clay nanocomposites prepared by *in situ* grafting-intercalating in melt. *Compositer Science and Technology*, **64** (2004), 1383–1389.

[17] M. J. Chung, L. W. Jang, J. H. Shim, and J.-S. Yoon. Preparation and characterization of maleic anhydride-g-polypropylene/diamine-modified clay nanocomposites. *Journal of Applied Polymer Science*, **95** (2005), 307–311.

[18] L. Cui and D. R. Paul. Evaluation of amine functionalized polypropylenes as compatibilizers for polypropylene nanocomposites. *Polymer*, **48** (2007), 1632–1640.

[19] Y. Wang, F.-B. Chen, and K.-C. Wu. Twin-screw extrusion compounding of polypropylene/organoclay nanocomposites modified by maleated polypropylenes. *Journal Applied Polymer Science*, **93** (2004), 100–112.

[20] A. Leuteritz, D. Pospiech, B. Kretzschmar, M. Willeke, D. Jehnichen, U. Jentzsch, and A. Janke. Properties of polypropylene clay nanocomposites modified with difunctional compounds. *Composite Interfaces*, **12**:3–4 (2005), 231–242.

[21] K. Wang, S. Liang, R. Du, Q. Zhang, and Q. Fu. The interplay of thermodynamics and shear on the dispersion of polymer nanocomposite. *Polymer*, **45** (2004), 7953–7960.

[22] M. Modesti, A. Lorenzetti, D. Bon, and S. Besco. Effect of processing conditions on morphology and mechanical properties of compatibilized polypropylene nanocomposites. *Polymer*, **46** (2005), 10237–10245.

[23] H.-S. Lee, P. D. Fasulo, W. R. Rodgers, and D. R. Paul. TPO based nanocomposites. Part 1: morphology and mechanical properties. *Polymer*, **46** (2005), 11673–11689.

[24] H.-S. Lee, P. D. Fasulo, W. R. Rodgers, and D. R. Paul. TPO based nanocomposites. Part 2: thermal expansion behavior. *Polymer*, **47** (2006), 3528–3539.

[25] D. H. Kim, P. D. Fasulo, W. R. Rodgers, and D. R. Paul. Effect of the ratio of maleated polypropylene to organoclay on the structure and properties of TPO-based nanocomposites. Part I: morphology and mechanical properties. *Polymer* **48**, (2007), 5960–5978.

[26] D. H. Kim, P. D. Fasulo, W. R. Rodgers, and D. R. Paul. Effect of the ratio of maleated polypropylene to organoclay on the structure and properties of TPO-based nanocomposites. Part II: thermal expansion behavior. *Polymer*, **49** (2008), 2492–2506.

[27] D. M. Delozier, R. A. Orwoll, J. F. Cahoon, N. J. Johnson, J. G. Smith Jr, and J. W. Connell. Preparation and characterization of polyimide/organoclay nanocomposites. *Polymer*, **43** (2002), 813–822.

[28] D. M. Delozier, R. A. Orwoll, J. F. Cahoon, J. S. Ladislaw, J. G. Smith Jr, and J. W. Connell. Polyimide nanocomposites prepared from high-temperature, reduced charge organoclays. *Polymer*, **44** (2003), 2231–2241.

[29] D. M. Delozier, R. A. Orwoll, J. F. Cahoon, J. S. Ladislaw, J. G. Smith Jr, and J. W. Connell. Polyimide nanocomposites prepared with a novel aromatic surfactant. *High Performance Polymers*, **15** (2003), 329–346.

[30] D. M. Delozier and D. C. Working. Polyetherimide/montmorillonite nanocomposites via *in-situ* polymerization followed by melt processing. *High Performance Polymers*, **16** (2004), 597–609.

[31] H. A. Stretz and D. R. Paul. Properties and morphology of nanocomposites based on styrenic polymers. Part II: effects of maleic anhydride units. *Polymer*, **47** (2006), 8527–8535.

[32] H. A. Stretz, D. R. Paul, R. Li, H. Keskkula, and P. E. Cassidy. Intercalation and exfoliation relationships in melt-processed poly(styrene-co-acrylonitrile)/montmorillonite nanocomposites. *Polymer*, **46** (2005), 2621–2637.

[33] H. A. Stretz, D. R. Paul, and P. E. Cassidy. Poly(styrene-co-acrylonitrile)/montmorillonite organoclay mixtures: a model system for ABS nanocomposites. *Polymer*, **46** (2005), 3818–3830.

[34] H. A. Stretz and D. R. Paul. Properties and morphology of nanocomposites based on styrenic polymers. Part I: styrene-acrylonitrile copolymers. *Polymer*, **47** (2006), 8123–8136.

[35] A. B. Morgan and J. D. Harris. Exfoliated polystyrene–clay nanocomposites synthesized by solvent blending with sonication. *Polymer*, **45** (2004), 8695–8703.

[36] J. W. Gilman, W. H. Awad, R. D. Davis, J. Shields, R. H. Harris Jr, C. Davis, A. B. Morgan, T. E. Sutto, J. Callahan, P. C. Trulove, and H. C. DeLong. Polymer/layered silicate nanocomposites from thermally stable trialkyimidazolium-treated montmorillonite. *Chemistry of Materials*, **14** (2002), 2776–3785.

[37] A. Samakande, P. C. Hartmann, V. Cloete, and R. D. Sanderson. Use of acrylic based surfmers for the preparation of exfoliated polystyrene–clay nanocomposites. *Polymer*, **48** (2007), 1490–1499.

[38] A. Samakande, P. C. Hartman, and R. D. Sanderson. Synthesis and characterization of new cationic quaternary ammonium polymerizable surfactants. *Journal of Colloid and Interface Science*, **296** (2006), 316–323.

[39] D. Shah, P. Maiti, E. Gunn, D. F. Schmidt, D. D. Jiang, C. A. Batt, and E. P. Giannelis. Dramatic enhancements in toughness of polyvinylidene fluoride nanocomposites via nanoclay-directed crystal structure and morphology. *Advanced Materials*, **16**:14 (2004), 1173–1177.

[40] D. Shah, P. Maiti, D. D. Jiang, C. A. Batt, and E. P. Giannelis. Effect of nanoparticle mobility on toughness of polymer nanocomposites. *Advanced Materials*, **175** (2005), 525–528.

[41] Y. Kim and J. L. White. Melt–intercalation nanocomposites with fluorinated polymers and a correlation for nanocomposite formation. *Journal of Applied Polymer Science*, **92** (2004), 1061–1071.

[42] Y. Kim and J. L. White. Modeling of polymer–clay nanocomposite formation. *Journal of Applied Polymer Science*, **101** (2006), 1657–1663.

[43] L. T. Vo and E. P. Giannelis. Compatibilizing poly(vinylidene fluoride)/nylon-6 blends with nanoclay. *Macromolecules*, **40**:23 (2007), 8271–8276.

[44] Y. Kim and J. L. White. Melt–intercalation nanocomposites with chlorinated polymers. *Journal of Applied Polymer Science*, **90** (2003), 1581–1588.

[45] G. W. Beall, C. E. Powell, J. Hancock, M. Kindinger, H. R. McKenzie, A. V. Bray, and C. J. Booth. Physical properties of CBDO based co-polyterephthalate nanocomposites. *Applied Clay Science*, **37** (2007), 295–306.

[46] M. Kracalik, M. Studenovsky, J. Mikesova, A. Sikora, R. Thomann, C. Friedrich, I. Fortelny, and J. Simonik. Recycled PET nanocomposites improved by silanization of organoclays. *Journal of Applied Polymer Science*, **106** (2007), 926–937.
[47] E. Pollet, C. Delcourt, M. Alexandre, and P. Dubois. Transesterification catalysts to improve clay exfoliation in synthetic biodegradable polyester nanocomposites. *European Polymer Journal*, **42** (2006), 1330–1341.
[48] C.-J. You, J.-G. Xu, S. Xi, X.-X. Duan, J. Shen, and D.-M. Jia. Properties and morphology of unsaturated polyester/acrylate-terminated polyurethane/organo-montmorillonite nanocomposites. *Chinese Journal of Polymer Science*, **235** (2005), 471–478.
[49] M. H. Beheshty, M. Vafayan, and M. Poorabdollah. Low profile unsaturated polyester resin–clay nanocomposite properties. *Polymer Compositer*, **305** (2009), 629–638.
[50] T.-M. Wu and C.-Y. Liu. Poly(ethylene 2,6-naphthalate/layered silicate nanocomposites: fabrication, crystallization behavior and properties. *Polymer*, **46** (2005), 5621–5629.
[51] R. K. Bharadwaj, A. R. Mehrabi, C. Hamilton, C. Trujillo, M. Murga, R. Fan, A. Chavira, and A. K. Thompson. Structure–property relationships in cross-linked polyester–clay nanocomposites. *Polymer*, **43** (2002), 3699–3705.
[52] Y. Someya and M. Shibata. Morphology, thermal, and mechanical properties of vinylester resin nanocomposites with various organo-modified montmorillonites. *Polymer Engineering and Science*, **44**:11 (2004), 2041–2046.
[53] News Report; Nanotechnology offers benefits for personal watercraft. *Reinforced Plastics*, **523** (2008), 8.
[54] C.-J. You, J.-G. Xu, S. Xi, X.-X. Duan, J. Shen, and D.-M. Jia. Properties and morphology of unsaturated polyester/acrylate-terminated polyurethane/organo-montmorillonite nanocomposites. *Chinese Journal of Polymer Science*, **235** (2005), 471–478.
[55] X. Zhang, R. Xu, Z. Wu, and C. Zhou. The synthesis and characterization of polyurethane/clay nanocomposites. *Polymer International*, **52** (2003), 790–794.
[56] P. K. Maji, P. K. Guchhait, and A. K. Bohowmick. Effect of nanoclays on the physic–mechanical properties and adhesion of polyester-based polyurethane nanocomposites: structure–property correlations. *Journal of Materials Science*, **44** (2009), 5861–5871.
[57] A. K. Kaushik, P. Podsiadlo, M. Qin, C. M. Shaw, A. M. Waas, N. A. Kotov, and E. M. Arruda. The role of nanoparticle layer separation in the finite deformation response of layered polyurethane–clay nanoparticles. *Macromolecules*, **42** (2009), 6588–6595.
[58] P. K. Maji, P. K. Guchhait, and A. K. Bhowmick. Effect of the microstructure of a hyperbranched polymer and nanoclay loading on the morphology and properties of novel polyurethane nanocomposites. *Applied Materials Interfaces*, **12** (2009), 289–300.
[59] C. E. Powell. Preparation of rubber compositions with organoclays. *US Patent* 2005/0090584 A1, 2005.

[60] S. Sadhu and A. K. Bhowmick. Preparation and properties of nanocomposites based on acrylonitrile-butadiene rubber, styrene-butadiene rubber, and polybutadiene rubber. *Journal Polymer Science: Part B: Polymer Physics*, **42** (2004), 1573–1585.

[61] M. I. Knudson Jr and C. E. Powell. Preparation of polymer nanocomposites by dispersion destabilization. *US Patent* 6849680 B2, 2005.

[62] Q.-X. Jia, Y.-P. Wu, Y.-L. Xu, H.-H. Mao, and L.-Q. Zhang. Combining in-situ organic modification of montmorillonite and the latex compounding method to prepare high-performance rubber–montmorillonite nanocomposites. *Macromolecular Materials and Engineering*, **291** (2006), 218–226.

8 Flame retardancy

An early observation by Blumstein [1] indicated that montmorillonite present in the polymerization of methyl methacrylate to produce polymer–clay composites significantly increased the thermal stability of the methyl methacrylate polymer in relation to polymethyl methacrylate prepared without the montmorillonite present. The polymer within the galleries of the montmorillonite was reported to have significantly higher thermal stability. Speculation on the cause of this enhanced thermal stability focused on restricted polymer chain mobility in the galleries and the prevention of oxygen diffusion into the galleries. The presence of oxygen during the thermal degradation of polymer–clay nanocomposites will be demonstrated to be a significant independent variable relating to the thermal degradation.

Little further activity is found in the literature until the advent of the importance of exfoliated layered clays in the dramatic enhancement of polymer mechanical performance at low concentrations was reported [2]. Subsequent systematic evaluations of the thermal stability of polymer–clay nanocomposites were initiated by Jeff Gilman's group at NIST and Emmanuel Giannelis' group at Cornell, with remarkable results. This work led to a dramatic increase in scientific investigations focused on the structure–property relationships of polymer–clay nanocomposites to thermal stability and flame retardancy.

An excellent review of the work on the flame retardancy of polymer nanocomposites was published in 2007 [3]. This chapter will focus on the evaluation of the proposed mechanisms for enhanced thermal stability of polymer–clay nanocomposites, the proposed relationships between enhanced thermal stability of polymer–clay nanocomposites and flame retardancy, and the synergies that develop between traditional flame retardants for polymers and polymer–clay nanocomposites.

8.1 Enhanced thermal stablility provided by polymer–clay nanocomposites

A large volume of published information is available [3] that demonstrates the significant enhancement of thermal stability of polymers when layered silicates are exfoliated to varying degrees into polymers. One of the observations found in this literature is an onset of thermal degradation of the polymer–clay nanocomposites below the thermal degradation temperature of the polymer without the presence of the clay. The majority of the thermal degradation evaluations are accomplished with the utilization of thermal gravimetric analysis (TGA). Speculation on the mechanism of this lower temperature degradation has focused on the quaternary ammonium ions that are exchanged onto, in many instances, montmorillonite that is employed to prepare the polymer–clay nanocomposite. The Hoffman elimination mechanism has been proposed as the chemical pathway for the degradation of the quat [4].

The features of the Hoffman elimination reaction mechanism are found in organic chemistry textbooks utilized to teach undergraduate organic chemistry. The first step in the reaction is to replace the halide counterion for the tetra alkyl quaternary ammonium ion with hydroxide ion by employing silver oxide slurry in water. Commercially available tetra alkyl quaternary ammonium ions are prepared by the reaction of a trialkyl amine with methyl chloride or dimethyl sulfate as alkylating agents. Benzyl quats are prepared by reacting a trialkyl amine with benzyl chloride. The counterion for the commercial quats, as a consequence, is usually chloride. The products of the first step in the Hoffman elimination reaction are the tetra alkyl quaternary ammonium ion reaction with a hydroxyl counterion and silver chloride. The next step is heating the hydroxyl counterion, tetra alkyl ammonium ion. The products of the next step are an amine, water, and an alkene. This elimination step proceeds through an E_2 mechanism. Generally, the least substituted alkene is produced. This is in violation of Zaitsev's rule which predicts the formation of the most substituted double bond for the alkene. The hydroxyl group functions as the nucleophile that abstracts a proton from the β carbon next to the ammonium ion. The amine group is the leaving group. This step proceeds in a concerted fashion. A strong base (the hydroxyl ion) is necessary to facilitate the elimination step. The base strength of the preexisting halide counterion for the quat is not strong enough.

Organomontmorillonites are prepared by exchanging tetra alkyl quaternary ammonium ions onto the surface of the clay. The existing counterion of the montmorillonite is a metal ion (sodium, potassium, etc.). The positive charged quat displaces the positively charged metal ion to produce the organoclay and the metal halide salt (sodium chloride, potassium chloride, etc.). The salt is washed away from the organoclay during the production of the organoclay. Precise amounts of quat are utilized in the process to insure that complete exchange occurs. The exchange capacity of the clay is determined to provide the information for the precise amount of quat needed for full exchange. The clay now serves the purpose of the counterion necessary for the E_2 elimination step in the proposed Hoffman elimination reaction of organoclays. Recent work focused on the acid-based properties of clays [5,6,7,8] indicates that the surface of montmorillonite as a counterion for the quat is potentially not basic enough to provide for the successful nucleophilic attack necessary for the E_2 step of the Hoffman elimination reaction. A particular revealing observation indicates that the ζ potential for montmorillonite with sodium counterions in water maintains a significantly negative value down to at least a pH of 2. Hence, the surface of montmorillonite is not basic enough to associate with protons at relatively high hydrogen ion concentrations to neutralize its surface charge.

An alternative mechanism for the thermal degradation of the quat exchanged onto the surface of montmorillonite could be reverse alkylation. The quats are prepared commercially by alkylation of trialkyl amines with alkyl halides. When the quats are exchanged at the capacity of the montmorillonite, the resulting metal halide salt is removed by washing the organoclay with water in a commercial process. Water could possibly serve as the nucleophile in the substitution reaction for the reverse alkylation at an elevated temperature to produce an amine, a proton, and alcohol. The proton would be expected to associate with the strongest base present, the amine, to produce an ammonium ion. The ammonium ion could serve as a counterion for the negatively charged montmorillonite. The charged head group of the quaternary ammonium ion is hydrophilic. The hydration energy of the association of water with the quat head group can be significant. Because the molecular weight of water is low (18 g/mol), a small amount of water by weight will provide a significant number of molecules for reaction when compared to the number of molecules present of the high molecular weight quats.

Thermal analysis of hexadecyltrimethyl quaternary ammonium ion exchanged onto montmorillonite [9] indicates that water could be present in the quat-clay gallery up to 130 °C. Reviews of published studies [3] of the thermal stability of organoclays indicate that water could be present up to 200 °C. These studies indicate that the addition of quat to montmorillonite past the exchange capacity compromises the thermal stability of the organoclay. Removal of quat from montmorillonite by solvent extraction increases the thermal stability of the organoclay [4,10]. The presence of halogen counterions associated with the unexchanged quaternary ammonium ions could allow for the reverse alkylation of the quat with halogens. There is always some residual chloride left behind in commercial organoclays owing to the difficulty in washing low permeability filter cakes. Analysis of the decomposition products of the thermal decomposition of organoclays generally indicates the presence of olefins, alkyl halides, aldehydes, and amines [2].

Solid state NMR analysis of nylon-6–clay nanocomposite that utilized dimethyl, dehydrogenated tallow quaternary ammonium ion as the surface modifier for the clay indicated that the quat decomposed during melt blending with the polymer to the methyl, ditallow amine [11]. No further decomposition of the amine occurred. Thermal dehydration of alcohols could produce alkenes; thermal oxidation of primary alcohols would produce aldehydes. Thermal stability of organomontmorillonites prepared with octadecyltrimethyl quaternary ammonium ions at very low exchange capacity (0.2 to 4.0) [12] indicates a relationship of exchange capacity and the loss of surfactant. The organoclays at the low range of exchange (0.2 and 0.4) have a mass loss at approximately 380°C. The boiling point of the surfactant is approximately 200°C. The organoclays exchanged at 0.6 and 0.8 have mass losses at 380°C and 280°C. The organoclays at the higher exchange (1.0 to 4.0) have mass losses at 380°C, 280°C, and 180°C. Presumably, the presence of hydrated sodium counterions associated with the montmorillonite can alter the stability of the quats exchanged onto the surface of the montmorillonite. This observation could support the hypothesis for the reverse alkylation of quats by water as the nucleophile. Sodium would be expected to have higher hydration energy than the head group of the quat. Hence, the sodium ions could preclude the association of water at the quat head group and slow the reverse alkylation step.

The thermal stability of commercial quats that are employed to prepare organoclays utilized in the preparation of polymer–clay nanocomposites are expected to be lower than the polymers in the polymer–clay nanocomposites. Commercial quats are designed for use in surfactant products (soaps, detergents, shampoos, etc.). The commercial quats are derived from fats and oils. Thermal degradation considerations are not major factors in the markets they serve. Fats and oils are utilized in food preparation, but one must be cognizant of their thermal degradation properties in order to ensure that no unfortunate flammability issues occur in the kitchen. Exchanging surface treatments onto montmorillonite that are designed for increased thermal stability have been very successful in increasing the thermal stability of organoclays [3].

A recent publication [13] illustrates the value of designing surface treatments for montmorillonite that are expressly developed for the preparation of organoclays with thermal stabilities commensurate or superior to the thermal stabilities of polymers utilized to prepare polymer–clay nanocomposites. The merging of oligomeric silsesquioxanes (POSS) with imidazolium charged head groups provides a very impressive thermally stable organomontmorillonite.

Several successful strategies are available in the literature [14] that increase the thermal stability of organic molecules. Full utilization of these strategies for the preparation of surface treatments of layered silicates with enhanced thermal stability for the development of polymer–clay nanocomposites has yet to be realized. An example of an effective strategy is the utilization of quaternary ammonium and phosphonium functional polystyrene as a surface treatment for montmorillonite that is employed to prepare polymer–clay nanocomposites [15]. TGA indicated a significant increase in the thermal stability of the organoclay and the polymer–clay nanocomposite. Imidazolium functional surface modifier for montmorillonite demonstrated a significant increase in the thermal stability of ABS terpolymer–clay nanocomposite when compared to the pure polymer and polymer–clay nanocomposites where the surface modification of the montmorillonite was produced with traditional quats [16]. These experiments were via TGA measurements.

The second significant independent variable that layered silicates provide to increase thermal stability of the polymer in polymer–clay nanocomposites is an increase of the melt viscosity. If thermal degradation of the polymer is diffusion controlled, an increase in viscosity of the polymer melt will slow the mass loss associated with gas escaping from the composite during TGA evaluations. The increase in viscosity of dispersions is a function of the surface area of the dispersed phase. For example, water-based dispersions will increase in viscosity as the particle size of the dispersed phase decreases at constant total volume of the dispersed phase. This is the result of an increase in total surface area of the dispersed phase. Particle–particle interaction has increased as a function of increased total particle surface area. The surface area [17] of fully exfoliated montmorillonite is approximately 750 m^2/g. This enormous number results in a significant increase in polymer–montmorillonite melt viscosity at low concentration of montmorillonite and low shear rates [18] when compared to the pure polymer. Some investigators utilize the melt viscosity of the polymer–montmorillonite composite as an indicator of the degree of exfoliation [19]. The montmorillonite concentration will continue to increase as the polymer degrades. This will further increase the viscosity of the polymer melt and slow gas evolution. This phenomenon is particularly noticeable in vertical polymer flammability tests that evaluate polymer melt dripping. Dripping is significantly decreased. These tests will be discussed later in this chapter. This behavior is favored with complete exfoliation of the montmorillonite which will maximize surface area. Intercalated montmorillonite in composites will be less effective in increasing the viscosity of the polymer melt.

The third significant independent variable that increases the thermal stability of polymer–clay nanocomposites is the thermal insulation properties. Heat flow through material is a function of the temperature difference in the direction of the heat flow, the area of the material exposed to the temperature difference, and the length of the material in which the heat is flowing. Thermal conductivity measurements of montmorillonite–polyurethane polymer nanocomposites [20] and polyethylene–Laponite (a synthetic hectorite) polymer nanocomposites [21] indicate a decrease in thermal conductivity as a function of increasing clay concentration in the polymer. The thermal conductivity of clay is higher than polymers (1 $Wm^{-1}K^{-1}$) [22]. The Maxwell thermal conductivity prediction [23] indicates a decrease in thermal conductivity with increasing levels of clay. Models have been developed that attempt to explain this phenomenon by an increase in thermal resistance at the interface between the polymer and nanoparticles [24,25,26]. The morphology of the nanoparticles (plate morphology for montmorillonite) appears to be a significant independent variable. Some of the more dramatic examples of the insulation performance of montmorillonite are in the development of ablative coatings for rockets [27,28,29]. The difference between the high temperatures from intense combustion on the front side of the ablative coating compared to the temperature on the back side of the coating is dramatically reduced when comparing the performances of the polymer–montmorillonite nanocomposite coating to the coating without montmorillonite. Thermal conductivity decreases as the concentration of montmorillonite is increased in the ablative coatings.

The fourth significant independent variable that increases the thermal stability of polymer–clay nanocomposites is the barrier properties provided by the anisotropic morphology of the clay particles [3]. The mechanism of barrier performance of montmorillonite is fully developed here in chapter 4 on "Barrier Properties." The barrier performance of montmorillonite prevents the diffusion of oxygen into the polymer composite and diffusion of vaporized organic molecules out of the polymer nanocomposite. The importance of the barrier to oxygen diffusion has been determined by TGA evaluations on the polymer–clay nanocomposites in inert atmospheres compared with oxygen atmospheres [3].

The orientation of the sample in the TGA apparatus is critical. Samples that have the montmorillonite particles fully exfoliated and arranged parallel with the surface of the sample will provide the maximum barrier performance. The surfaces of montmorillonite are minimally 150 nm to 200 nm across. The edges of the particles are approximately 1 nm across. If the edges of the particles are aligned with the surface of the sample that is evaluated by TGA, the barrier protection of the montmorillonite will be minimal. This is a factor in the difference in cone calorimetry evaluations of polymer–clay nanocomposite panels for flame retardancy with the montmorillonite plate orientation parallel to the surface of the sample and vertical flame retardancy evaluations of the same samples where the particle edges provide virtually no flame retardancy.

Enhanced barrier performance is provided by montmorillonite as the polymer is degraded. The concentration of montmorillonite increases as a function of the degradation of the polymer. This provides added barrier protection. The polymer is in a melt form when thermal degradation occurs. Evaluations of the surface of the polymer–clay nanocomposite samples as a function of thermal degradation indicate that the montmorillonite is concentrated further by a percolation mechanism, presumably by the gas escaping from the sample and forcing the montmorillonite particles to the surface [2]. Analysis of the change in montmorillonite concentration as a function of temperature for polypropylene–clay nanocomposites supports this mechanism [30].

The fifth significant independent variable that increases the thermal stability of polylmer–clay nanocomposites is the alteration of the kinetics for the thermal degradation of the polymer. Blumstein [1] observed that poly(methyl methacrylate) present in the galleries between the montmorillonite in the polymer composite appeared to have greater thermal stability than the polymer outside the galleries. An evaluation of this effect is an examination of the rate-determining step(s) for the thermal degradation of the polymer.

The Arrhenius equation is a relationship between temperature and the change in reaction rate. Arrhenius established experimentally that the change in reaction rate (k) can be a function of the change in the temperature at which the reaction occurs.

$$\frac{\delta \, ln \, k}{\delta T} = \frac{E_a}{RT^2}, \tag{8.1}$$

where k is the observed rate of reaction, T is the temperature of the reaction in degrees Kelvin, R is the gas constant, and E_a is the energy of activation (the energy necessary for a chemical reaction to reach the transition state). Rearranging Equation 8.1 by multiplying both sides by δT gives:

$$\delta \, ln \, k = \frac{E_a \, \delta T}{RT^2}. \tag{8.2}$$

Integrating both sides of Equation 8.2 above,

$$\int \delta \, ln \, k = \int \frac{E_a \, \delta T}{RT^2} \tag{8.3}$$

yields:

$$\ln k = \frac{-E_a}{RT} + \text{integration constant.} \tag{8.4}$$

Taking the antinatural log of both sides of the equation yields

$$k = \frac{-E_a}{Ae^{RT}}, \tag{8.5}$$

where A is the integration constant.

Utilizing this general approach, the activation energy for the thermal degradation of ultrahigh molecular weight polyethylene–montmorillonite polymer composites determined by TGA increased when compared to the pure polymer [31] at relatively low concentrations of montmorillonite (0.5 and 1.0 wt.%).

An adaptation of the Arrhenius equation referred to as the Kissinger method [32] was employed for this analysis. Determination of the activation energy for thermal degradation of polycarbonate–acrylonitrile-butadiene-styrene copolymer–montmorillonite composite when compared to the polymer blend without montmorillonite provided a similar trend [33]. The Arrhenius adaptation for this analysis was the Freeman–Carroll [34] method. These evaluations were also done by TGA. The concentration of montmorillonite was 5 wt.%. The TGAs were run with nitrogen and air. The activation energies for the TGA analysis in nitrogen were significantly higher than the activation energies determined in air (when oxygen was present). Analysis of the data indicated that the mechanism for thermal degradation of the polymer–clay composite in air could be significantly different than thermal decomposition in a nitrogen atmosphere.

The activation energy for the thermal degradation of polypropylene–clay nanocomposites by TGA was also determined [35]. Activation energies were determined with a nitrogen atmosphere. No evaluations were done with air. The results obtained with the Kissinger method, the Freeman–Carroll method, and the Coats–Redfern method [36] were very comparable and consistent with each other. The activation energies were determined with polymer–clay composites of different degrees of montmorillonite exfoliation. The concentration of montmorillonite in the polypropylene composites was 5 wt.%. The energy of activation increased with increased level of clay exfoliation. The thermal degradation of the polymer was consistent with first-order kinetics; the exfoliated montmorillonite present in the polymer provided a zero-order kinetic model for the thermal degradation.

The implication from the above work is that polymer present in the galleries of montmorillonite through intercalation and exfoliation has enhanced thermal stability provided by a different thermal degradation mechanism when compared to the pure polymer. Calculations of the activation energy of polyurethane-imide–clay nanocomposites [37] utilizing the Broido [38] and Coats–Redfern equations were consistent with a mechanism that increased the thermal stability of the polymer–clay nanocomposite in relation to the pure polymer. An optimization of melt-blending processing that improved the exfoliation efficiency of polymer–clay nanocomposites resulted in increased thermal stability by TGA when compared to polymer–clay nanocomposites with inferior exfoliation, and pure polymer [39].

Analysis of the thermal degradation products indicates that the mechanism of polymer degradation has been altered when clay is present in the polymer [40]. This behavior supports the argument that the change in the energy of activation is indicative of the alteration of the thermal degradation mechanism for polymer–clay nanocomposites when compared to the pure polymer. The analysis also indicates that the mechanism is a function of polymer type. For example, char formation during TGA evaluations of poly(vinyl chloride)/montmorillonite polymer nanocomposites is significantly greater than the pure polymer [41]. This is a significant variable in the measurement of the thermal stability of the polymer nanocomposite.

Polyaniline–montmorillonite nanocomposites were prepared without surface modification of montmorillonite [42]. The polymer nanocomposite was significantly more thermally stable when compared to the pure polymer. Ethylene-vinyl acetate copolymer–clay nanocomposites apparently thermally degrade by a different mechanism than the pure polymer [43]. These observations are consistent with the above thesis.

8.2 Relationships between enhanced thermal stability of polymer–clay nanocomposites and flame retardancy

Two features must be included in this discussion to provide a meaningful discussion of the relationship between enhanced thermal stability of polymer–clay nanocomposites and flame retardancy. A cogent definition of flame retardants needs to be identified. Two definitions are discussed by Bartholmai and Schartel [19]:

1. A fire retardant is anything that delays the start or decreases the propagation of a fire, whatever the size of the effect.
2. A fire retardant is a commercial product which improves a polymer material so that it can pass fire regulations and fire tests without too much influence of other properties.

These definitions will be utilized to judge the flame-retardant effectiveness of clay as a dispersed phase in polymers.

The second feature is the identification of a measuring system for the quantification of the flame-retardant effectiveness of clay as a dispersed phase in polymers. The cone calorimeter is an excellent tool for this purpose [3]. An additional device referred to as a simulated solid rocket motor (SSRM) will be introduced. The SSRM is designed to evaluate polymer composite flame retardancy at extreme conditions.

The development of the cone calorimeter was reported by the National Bureau of Standards [44] (now NIST) in 1982. Further refinements were reported [45] in 1987. The features of the cone calorimeter are reported in depth [46]. The features of the cone calorimeter consist of a stage where the sample is mounted. The stage is attached to a load cell which measures the change in sample weight as a function of the time of combustion. The sample stage can also be fitted with a thermocouple to follow the change in temperature on the back side of the sample as it combusts. An upside-down cone is fitted above the sample at a prescribed distance. Heating elements are present inside the cone for heating the sample at a measured rate for combustion. A spark igniter is positioned near the sample to provide an ignition source. An exhaust hood is positioned over the cone heater. Soot collection, temperature of the combustion gases, samples for analysis of the combustion gases, pressure measurements, and laser photometry for smoke measurements are accomplished by sampling from the duct of the exhaust hood. The information that is provided from the cone calorimeter test is:

- time to ignition;
- total heat released;
- maximum heat release rate;
- average heat release rate;
- effective heat of combustion;
- average smoke production;
- production of CO;
- mass loss.

The results are reported as a function of the size of the sample surface in square meters. It is significant that the rate of heating the sample by the cone calorimeter is much faster than the rate of heating by TGA reported above for the thermal stability evaluations.

The observation that polymer–clay nanocomposites have significantly lower peak heat release rates (PHRRs) when compared to the pure polymer [47] stimulated a dramatic effort focused on the evaluation of the flame-retardant potential of clay dispersed in polymer. The decrease in PHRR can be related directly to the decrease in the spread of fire from one combustible material to another. This affect is directly applicable to definition (1) above. Figures 8.1 and 8.2 contain a comparison of cone calorimeter results for a pure polymer and a nanocomposite of that polymer with montmorillonite.

A disappointment in the observations of the flame-retardant behavior of clay in polymers was the decrease in the time to ignition for polymer–clay nanocomposites when compared to pure polymer. The commercial quaternary ammonium ions that were utilized to modify the surface of the clay for improved exfoliation efficiency in the polymers studied are prepared from fats and oils. The presence of these quats in the polymer–clay nanocomposites is tantamount to throwing fat into the fire. The decrease in the time to ignition for the polymer–clay nanocomposites that contain quats prepared from fats and oils is expected. Removal of some of the surfactant from the clay by extraction resulted in a longer time to ignition time for the polymer–clay nanocomposite prepared with the clay with lower surfactant content [48].

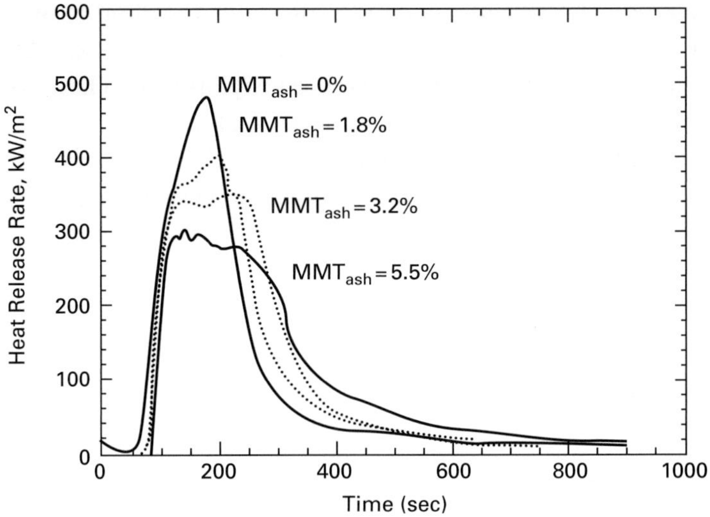

Figure 8.1 HRR of SAN/organomontmorillonite polymer nanocomposites as a function of montmorillonite content prepared from dimethyl, dehydrogenated tallow quaternary ammonium ion.

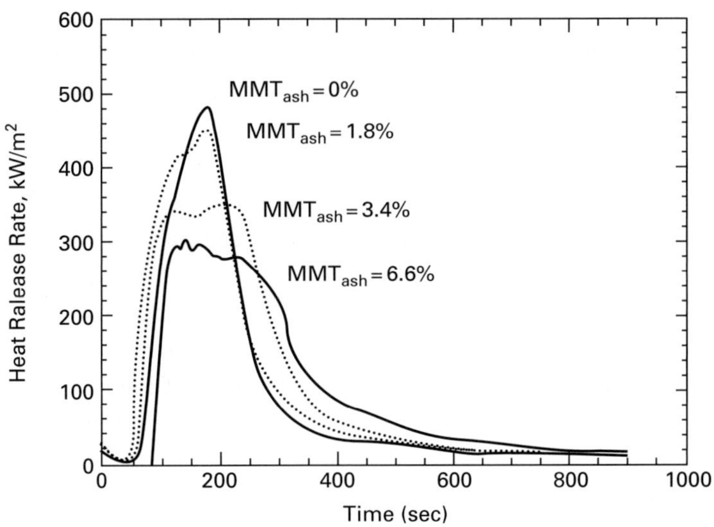

Figure 8.2 HRR of SAN/organomontmorillonite polymer nanocomposites as a function of montmorillonite content prepared from dimethyl, hydrogenated tallow ammonium ion.

The observation of decreasing time to ignition as a function of organoclay content in the polymer compromises the classification of organoclays as a flame retardant based on definition (1) above. Increasing the thermal stability of the surface treatment that is employed to expedite the exfoliation of clay into polymers alleviates the shorter time to ignition behavior. Progress in developing strategies for increased thermal stability of clay surface treatments has been significant. These strategies are discussed in the thermal stability section above.

The full complement of commercially available polymer types has been evaluated for flame retardancy as polymer–clay nanocomposites by cone calorimetry [3]. The type of polymer is a significant independent variable as regards the flame-retardant behavior of the polymer–clay nanocomposite. All of the significant independent variables that were discussed above that relate to thermal stability of the polymer–clay nanocomposites are applicable to flame-retardant behavior.

NIST organized a group of government and industry participants whose focus was a fundamental investigation to determine the mechanism of flame-retardant behavior of clay in polymer–clay nanocomposites [49]. The instrument of choice for the evaluations was the cone calorimeter. An important observation of this study was that polymers that normally burn without the development of a char (polystyrene, polypropylene, nylon, and ethylene-vinylacetate copolymer) develop a char when they are present in a polymer–clay nanocomposite. The mechanisms that are applicable in the increase in thermal stability for polymer–clay nanocomposites appear to be significant with an increase in flame retardancy through char formation. The polymers are prevented from thermally degrading completely to gases and form carbon residues. Examination of the char indicated a char–clay nanocomposite structure. The char–clay structure will retard the diffusion of gases produced in the polymer–clay through thermal decomposition. These gases are necessary for further combustion. The thickness of the char and its mechanical strength appear to be a function of the degree of exfoliation of the clay in the polymer and the concentration of clay in the polymer. Chars with polymers that have exfoliated clay and higher clay concentrations have a more integrated appearance and fewer cracks [48].

Polymer–clay nanocomposites that have the best fire-retardant performance evaluated by the cone calorimeter have clay particles oriented parallel to the surface of the sample. Vertical fire retardant tests (UL-94) of these samples do not demonstrate improved performance because the edges of the particles are exposed to the fire. The edges of clay are very thin (approximately 1 nm). Hence, the mechanisms predicated on barriers provided by the clay are not applicable. Mechanisms associated with increased melt viscosity are apparent with vertical fire-retardant testing. Dripping during the burning of the vertical samples is greatly reduced [3].

Cone calorimeter evaluations with and without the presence of oxygen seem to mimic the TGA results for thermal stability of polymer–clay nanocomposites [48]. Degradation of the polymer–clay nanocomposite to form a char–clay composite in the cone calorimeter in a nitrogen atmosphere is enhanced with oxygen present to further reduce the amount of char–clay composite at the surface of the sample. Presumably, the concentration of clay in the char is increased to a level during oxygen degradation where the diffusion of oxygen into the sample is compromised. This proposed mechanism is supported by work focused on the thermal oxidation of polypropylene–clay nanocomposites in the molten state [30,50].

The morphology of the clay appears to be a significant independent variable with regard to PHRR in cone-calorimetric evaluations of polymer–clay nanocomposites. A comparison of organomontmorillonite with organohectorite polymer nanocomposites with poly(methylmethacrylate) indicates that the organohectorite with the same quaternary ammonium ion modification as the organomontmorillonite has higher PHRR [37]. Hectorite has a lathlike structure as opposed to the sheetlike structure of montmorillonite. The exchange capacity of hectorites is usually about half that of montmorillonite. If barrier performance is a significant variable with regard to PHRR, the montmorillonite would be expected to be superior to hectorite.

A synthetic clay [47] produced by Dow Chemical Company demonstrated lower PHRR when compared to montmorillonite in ethylene-vinyl acetate copolymer–clay nanocomposites. The characterization of the synthetic clay is not available. The morphology difference between the synthetic clay and montmorillonite could provide further confirmation of the importance of clay morphology with regard to flame-retardant performance in the cone calorimeter.

The impact on the PHRR when substituting a polymer oligomer as a surface treatment on montmorillonite for a quaternary ammonium ion is not a panacea [51,52]. The comparison of methylmethacrylate oligomer-treated montmorillonite with quat-treated montmorillonite in poly(methylmethacrylate), polystyrene, high-impact polystyrene, acrylonitrile-butadiene-styrene terpolymer, polypropylene, and polyethylene nanocomposites indicated sensitivity to polymer type and a poor correlation to the degree of exfoliation determined by X-ray analysis and TEM. The impact of polymer structure associated with the montmorillonite appears to be a significant variable relating to the PHRR of these composites. In this book, chapter 4 on barriers addresses in depth polymer structure associated with montmorillonite. The PHRR is reduced for these polymer–clay composites as the concentration of montmorillonite is increased. A correlation between the degree of exfoliation of montmorillonite in these polymers and PHRR did not appear to be significant.

Other work with a traditional quat modification of montmorillonite and high-impact polystyrene is consistent with the above work in observing a decrease in PHRR as a function of increasing montmorillonite concentration in the polymer [53]. Work with styrene-acrylonitrile copolymer–clay nanocomposites with a traditional quat [54] surface modification of the montmorillonite demonstrated a significant decrease in PHRR as a function of montmorillonite concentration. The degree of exfoliation determined by solid state NMR was about 40% for this system.

Increasing the thermal stability of the quaternary ammonium surface treatment of montmorillonite provides for increased thermal stability of the polymer–clay nanocomposite and improved PHRR. Utilization of quinolinium and pyridinium head groups for the quats was very effective in this regard [55]. Biphenyl modification of the quaternary ammonium ion had a similar affect [56]. Quaternary ammonium ion functional polybutadiene as a surface treatment for clay also increases the thermal stability of the polymer–clay nanocomposites when the polymer was polystyrene, high-impact polystyrene, ABS terpolymer, polypropylene, poly(methylmethacrylate), and polyethylene [57]. However, the PHRR results are not encouraging. The quat functional polybutadienes utilized for the surface treatment of the clay apparently compromise the effective dispersion of the organoclays into the polymers listed above. The poor dispersion reflecting on minimal changes of PHRR supports the importance of the mechanisms outlined above for flame retardancy of polymer–clay nanocomposites. Carbazole modification of the quaternary ammonium ions utilized to modify the surface of montmorillonite [58] that was employed to prepare polystyrene–clay nanocomposites was not as effective at increasing the thermal stability of the polymer nanocomposite when compared to a traditional quat. However, when they were included in a bulk polymerization process to prepare the polymer–clay nanocomposite, the PHRR decreased significantly when compared to the pure polymer.

A SSRM provides a method to determine flame retardancy at more intense testing conditions. Figure 8.3 pictures the SSRM being utilized to test an ablative coating's performance. The features of the SSRM are the

Figure 8.3 SSRM in action.

combustion of oxygen and kerosene under pressure to produce a heat flux of up to 14000 Kw/m^2 at a temperature of 2200°C. The velocity of the gases from the SSRM will range up to Mach 7. The polymer–clay nanocomposites are mounted on a steel plate with a thermocouple attached to the back of the plate to measure the back side temperature during the test as a function of time. The samples are mounted 90° to the rocket blast. The surface temperatures are measured by an infrared pyrometer during the test. In addition to the temperature measurements, mass loss and dimension changes in the sample are measured after the test.

Significant improvements in lowering the back side temperature and erosion rate of the polymer composite are observed with the addition of montmorillonite to the polymer formulations [27,28,29] when compared to commercial ablative coating formulations. All of the mechanisms described above that pertain to enhanced thermal stability with polymer–clay nanocomposites are applicable to ablative coating performance. Figures 8.4 and 8.5 picture the results of commercial ablative coatings when tested with the SSRM. Figure 8.6 represents the improvement in ablative performance with a water-based polyurethane–montmorillonite nanocomposite with a loading of 15% and an added dispersion of carbon fiber at 3% loading. The measurements that

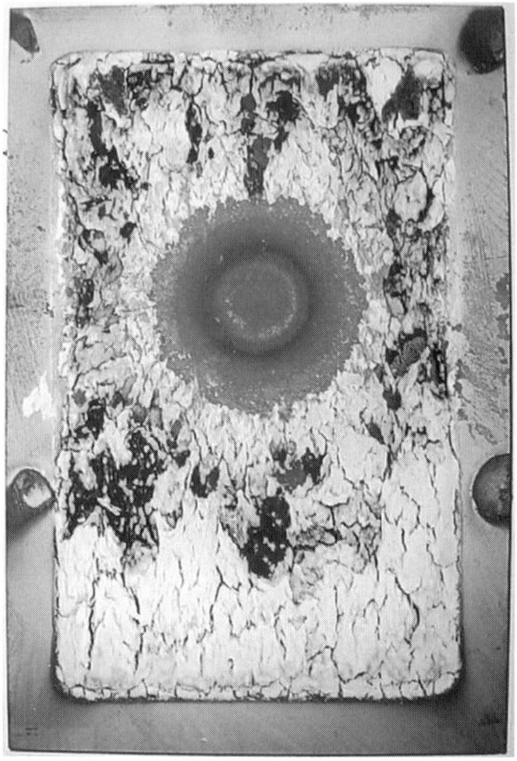

Figure 8.4 Aramid fiber-filled EPDM.

demonstrate the significant improvement in performance provided by the polymer–montmorillonite nanocomposite are summarized in Table 8.I.

Cone calorimetric evaluations of polymer–clay nanocomposites indicate that PHRR and mass loss rate (MLR) can be significantly reduced when compared to the pure polymer. However, in many cases, the ignition temperature is lower, the total heat released (THR) has not changed, and the total mass loss (TML) has not changed for the polymer–clay nanocomposites when compared to the pure polymer. An examination of the flame-retardant behavior of polymer–clay nanocomposites indicates that the presence of the clay delays the decomposition of polymer in the cone calorimeter test and does not prevent the decomposition. These observations in relation to the definitions listed above for flame retardants excludes clay from being considered to be a flame retardant in the same category as commercially available flame retardants. Because of these inadequacies, considerable effort has been made to identify synergies that may exist between commercial flame retardants and clay in polymer.

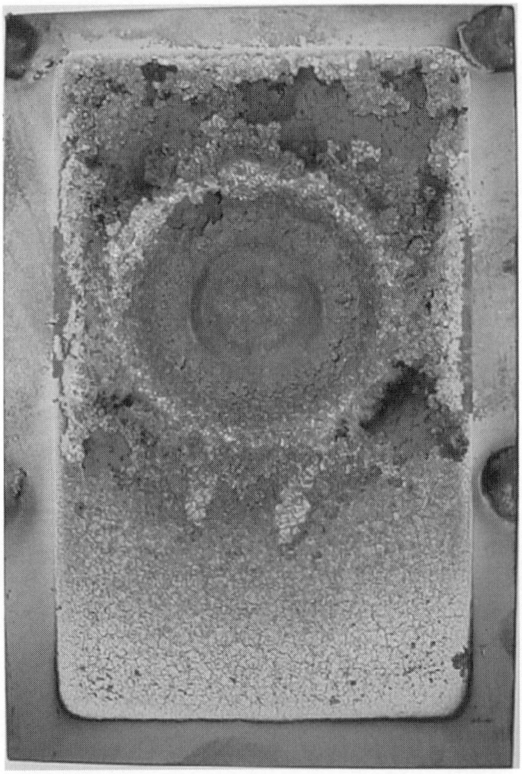

Figure 8.5 Silica-filled EPDM.

Figure 8.6 Bayhydrol 122 with 15% montmorillonite and 3% Hexcel carbon fiber.

Table 8.1 Ablative performance of the waterbased, polyurethane nanocomposite coating compared with the two commercial ablative rocket coatings.

Measurements	Aramid fiber-filled Epdm	Silica-filled Epdm	Bayhydrol 122 + 15% Montmorillonite + 3% Hexcel carbon fibers
Thickness of coating and steel before test in inches	0.242	0.213	0.685
Thickness of the steel before the test in inches	0.125	0.125	0.259
Thickness of the steel after the test in inches	0.125	0.097	0.259
Change in thickness of the coupon after the test in inches	—	– 0.036	—
Weight of steel before coating in grams	383.4		531.3
Weight of the coated steel before the test in grams	405.8	400.0	616.6
Weight of the coated steel after the test in grams	390.4	383.2	590.6
Weight change of the coating from testing in grams	−7.0		−26.0
% change in coating weight	−31.25		−30.48
Backside temperature in °C and the time to plateau in s.	119	157	74@75 seconds
Frontside temperature in °F	1974	2116	1888

8.3 Evaluations of potential synergies between traditional flame retardants for polymers and polymer–clay nanocomposites

An extensive survey of evaluations of the potential synergy between traditional flame retardants and polymer–clay nanocomposites is found in the literature [3]. Because the mechanisms of the traditional flame retardants vary, they will be categorized as a function of those mechanisms in this section.

Magnesium hydroxide and aluminum trihydrate are utilized as flame retardants in polymers. They function as flame retardants by releasing water during a fire. The water release cools the substrate and dilutes the combustible gases and oxygen that are required for the fire to continue. The concentration of magnesium hydroxide and aluminum trihydrate needs to be formulated in the polymer as high levels (up to 60%) for them to be effective fire retardants. This level of mineral content in a polymer seriously compromises the mechanical properties. Montmorillonite that is formulated into polymer has allowed for a lower concentration on magnesium hydroxide and aluminum trihydrate to be utilized, which has enhanced the mechanical properties of the polymer composite with a minimal sacrifice in mechanical properties [3,59,60,61,62,63,64]. The mechanisms that are applicable above for the increase in thermal stability for polymer–clay nanocomposites are also significant in polymer–clay nanocomposite formulations that contain magnesium hydroxide and aluminum trihydrate. The synergy of aluminum trihydrate formulated with polymer–clay nanocomposites has resulted in the commercial development of a coaxial cable that passes UL 1666 that has a formulation containing aluminum trihydrate and an ethylene-vinyl acetate copolymer–clay nanocomposite in the outer sheath [63].

Montmorillonite is a very anisotropic particle with sheet morphology. An isotropic form of magnesium hydroxide (magnesium hydroxide sulfate hydrate) has fiber morphology with diameters less than 1 μm and lengths longer that 10 μm. The combination of these anisotropic particles in polyethylene demonstrated a significant reduction in the PHRR and the mean MLR [65] at lower levels of magnesium hydroxide. The combination of these anisotropic particles could also enhance the mechanical performance of this polymer–clay nanocomposite.

The addition of intumescent materials to polymers has been demonstrated to be an effective approach to improve flame retardancy for polymers. The mechanism of the flame retardancy involves the formation of a char foam on the surface of a substrate during combustion. This char foam provides insulation of the substrate from the fire. The components of the intumescent fire retardant (IFR) package are ammonium polyphosphate, an alcohol, and a melamine. The ammonium polyphosphate decomposes to phosphoric acid during the combustion to produce phosphoric acid and ammonia. The phosphoric acid reacts with the alcohol to produce the char. The melamine is referred to as the blowing agent. Ammonia is produced from the melamine to produce the foaming. IFR packages compromise the mechanical performance of polymers and are more expensive than magnesium hydroxide and aluminum trihydrate.

Potential synergies were explored between polymer–clay nanocomposites and IFR packages to minimize the compromise of mechanical performance of the polymer and decrease the level of IFR needed for fire retardancy to increase cost effectiveness [3,64,65,66,67,68]. The theme through all of these studies was a synergistic effect of the clay to produce:

- more char at a faster rate;
- a reduction of PHRR;
- reduced smoke generation;
- a decrease in the THR;
- an increase in the residual mass upon combustion when compared to the utilization of the IFR in polymer without the clay.

All of the mechanisms discussed above with regard to the increase in thermal stability of polymer–clay nanocomposites are applicable to these results.

Halogenated flame retardants for polymers have been utilized extensively in the industry. Halogens presumably function in the gas phase of a fire to capture free radicals that are generated during combustion and thus minimize further combustion. Metal oxides (for example, antimony trioxide) are utilized with halogenated flame retardants to further the efficiency of this chemistry. For example, the antimony trioxide will react with halogenated flame retardants to produce antimony trihalides. Antimony trihalides are volatile and (as a consequence) participate in minimizing combustion in the gas phase through free radial reactions. Halogenated flame retardants produce toxic and corrosive gases during combustion. Polymer–clay nanocomposites have been evaluated with halogenated flame retardants in order to lower the amount of flame retardants necessary to maintain performance and mitigate the unfavorable effects of halogens in combustion products [3,69,70,71]. The clay provides improvement in the PHRR and the halogens provide a decrease in the THR. The synergy between the polymer–clay composites and halogens appears to be significant down to low levels of halogen content.

Red phosphorus and phosphates are also utilized commercially as halogen free flame retardants. The mechanism for their flame-retardant behavior is the production of a char at the surface of the substrate during combustion that provides a barrier for further combustion. The char formation is presumably the result of the production of phosphoric acid during combustion that degrades the polymer. Polymer–clay nanocomposites have been employed to further the flame-retardant effectiveness of phosphorus and improve the mechanical performance of the polymer composites [3,72]. The synergy between the phosphorus and the clay during combustion apparently decreases the PHRR, increases char formation, and decreases the peak MLR.

8.4 Summary and conclusions

The significant independent variables of polymer–clay nanocomposites that determine the increased thermal stability of the polymer are:

- the nature of the surface treatment of the clay;
- the increase in melt viscosity of the polymer by the clay;
- the decrease in thermal conductivity of the polymer–clay nanocomposites as a function of increased clay concentration;
- the increase in barrier properties of the composites with increasing clay content;
- the change in the thermal decomposition mechanism as demonstrated by the increase in the activation energy for thermal decomposition.

These variables that relate to increased thermal stability of the polymer in polymer–clay nanocomposites extrapolate directly to improved flame retardancy of polymer–clay nanocomposites. The improvement in flame retardancy provided by polymer–clay nanocomposites does not completely satisfy the criteria found in the definitions for traditional commercial flame retardants. Lower ignition temperatures, no change in the total heat release, and no change in total mass loss as measured by cone calorimetry of the polymer–clay nanocomposites when compared to the pure polymer prevent the classification of clay in polymer–clay nanocomposites as a flame retardant. Flame-retardant synergies between commercial flame retardants and polymer–clay nanocomposites have mitigated some of the flame-retardant deficiencies of the polymer–clay nanocomposites and allow for their commercial introduction into products.

Because of the multifaceted features of clay as a nanoparticle, the benefits in polymer–clay nanocomposites range from increased mechanical performance, barrier performance, and thermal stability. A systems approach to the design of polymer–clay nanocomposites with excellent flame retardancy will provide superior solutions in relation to formulating existing flame retardants with polymer–clay nanocomposites. Providing surface modifications for the clay with higher thermal stability that will not compromise the mechanical and barrier performance of the composites is essential. Integrating additional components into the polymer–clay nanocomposites that provide additional effective mechanisms for flame retardancy that are not provided by the clay and do not compromise mechanical and barrier performance will provide a uniquely value-added solution for the complex issue of enhanced flame retardancy for commercial products.

References

[1] A. Blumstein. Polymerization of adsorbed monolayers. II: thermal degradation of the inserted polymer. *Journal of Polymer Science: Part A*, **3** (1965), 2665–2672.

[2] A. Okada, M. Kawasumi, T. Kurauchi, and O. Kamigaito. Synthesis and characterization of a nylon 6–clay hybrid. *Polymer Preprints*, **28**:2 (1987), 447–448.

[3] A. B. Morgan and C. A. Wilkie, eds. *Flame Retardant Polymer Nancomposites* (Wiley-Interscience, Hoboken, NJ, 2007).

[4] W. Xie, Z. Gao, W.-P. Pan, D. Hunter, A. Singh, and R. Vaia. Thermal degradation chemistry of alkyl quaternary ammonium montmorillonite. *Chemistry of Materials*, **13** (2001), 2979–2990.

[5] M. Duc, F. Gaboriaud, and F. Thomas. Sensitivity of the acid-base properties of clays to the methods of preparation and measurement 1: literature review. *Journal of Colloid and Interface Science*, **289** (2005), 139–147.

[6] B. Tyagi, C. D. Chudasama, and R. V. Jasra. Characterization of surface acidity of an acid montmorillonite activated with hydrothermal, ultrasonic and microwave techniques. *Applied Clay Science*, **31** (2006), 16–28.

[7] M. Duc, F. Gaboriaud, and F. Thomas. Sensitivity of the acid-base properties of clays to the methods of preparation and measurement 2: evidence from continuous potentiometric titrations. *Journal of Colloid Interface Science*, **289** (2005), 148–156.

[8] P. Zarzycki and F. Thomas. Theoretical study of the acid-base properties of the montmorillonite/electrolyte interface: influence of the surface heterogeneity and ionic strength of the potentiometric titration curves. *Journal of Colloid Interface Science*, **302** (2006), 547–559.

[9] H. He, Z. Ding, J. Zhu, P. Yuan, Y. Xi, D. Yang, and R. L. Frost. Thermal characterization of surfactant-modified montmorillonites. *Clay and Clay Minerals*, **53** (2005), 287–293.

[10] H. He, J. Duchet, J. Galy, and J.-F. Gerard. Influence of cationic surfactant removal on the thermal stability of organoclays. *Journal of Colloid Interface Science*, **295** (2006), 202–208.

[11] D. L. VanderHart, A. Asano, and J. W. Gilman. Solid-state NMR investigation of paramagnetic nylon-6 clay nanocomposites. 2: measurement of clay dispersion, crystal stratification, and stability of organic mondifiers. *Chemical Materials*, **13** (2001), 3796–3809.

[12] Y. Xi, W. Martens, H. He, and R. L. Frost. Thermogravimetric analysis of organoclays intercalated with the surfactant octadecyltrimethylammonium bromide. *Journal of Thermal Analysis and Calorimetry*, **81** (2005), 91–97.

[13] D. M. Fox, P. H. Maupin, R. H. Harris Jr, J. W. Gilman, D. V. Eldred, D. Datsoulis, P. C. Trulove, and H. C. de Long. Use of a polyhdral oligomeric silsesquioxane (POSS)-imidazolium cation as an organic modifier for montmorillonite. *Langmuir*, **23** (2007), 7707–7714.

[14] G. L. Tullos and P. E. Cassidy. Polymers derived from hexafluoroacetone: 12F-poly(ether ketone). *Macromolecules*, **24** (1991), 6059–6064.

[15] B. N. Jang and C. A. Wilkie. The effects of clay on the thermal degradation behavior of poly(styrene-co-acrylonitrile). *Polymer*, **46** (2005), 9702–9713.

[16] M. Modesti, S. Besco, A. Lorenzetti, V. Causin, C. Marega, J. W. Gilman, D. M. Fox, P. C. Trulove, H. C. de Long, and M. Zammarano. ABS/clay nanocomposites obtained by a solution technique: influence on clay organic modifiers. *Polymer Degradation and Stability*, **92** (2007), 2206–2213.

[17] H. van Olphen. *An Introduction to Clay Colloid Chemistry for Clay Technologists, Geologists and Soil Scientists* (Interscience Publishers, New York, 1963).

[18] R. Wagener and T. J. G. Reisinger. A rheological method to compare the degree of exfoliatioin of nanocomposites. *Polymer*, **44** (2003), 7513–7518.

[19] M. Batholmai and B. Schartel. Layered silicate polymer nanocomposites: new approach or illusion for fire retardancy? Investigations of the potentials and the tasks using a model system. *Polymer Advanced Technologics*, **15** (2004), 355–364.

[20] K. J. Yao, M. Song, D. J. Hourston, and D. Z. Luo. Polymer/layered clay nanocomposites: 2 polyurethane nanocomposites. *Polymer*, **43** (2002), 1017–1020.

[21] S. H. Lee, J. E. Kim, H. H. Song, and S. W. Kim. Thermal properties of maleated polyethylene/layered silicate nanocomposites. *International Journal of Thermophysics*, **25** (2004), 1585–1595.

[22] N. H. Abu-Hamdeh and R. C. Reeder. Soil thermal conductivity: effects of density, moisture, salt concentration, and organic mattter. *Soil Science Society of America Journal*, **64** (2000), 1285–1290.

[23] J. C. Maxwell. *A Treatise on Electricity and Magnetism*, 3rd edn (Dover, New York, 1954).

[24] Y. Benveniste. Effective thermal conductivity of composites with a thermal contact resistance between the constituents: nondilute case. *Journal Applied Physics*, **61** (1987), 2840–2843.

[25] C.-W Nan, R. Birringer, D. R. Clarke, and H. Gleiter. Effective thermal conductivity of particulate composites with interfacial thermal resistance. *Journal of Applied Physics*, **81** (1997), 6692–6699.

[26] G. Chen. Particularities of heat conduction in nanostructures. *Journal of Nanoparticle Research*, **2** (2000), 199–204.

[27] R. A. Vaia, G. Price, P. N. Ruth, H. T. Nguyen, and J. Lichtenhan. Polymer/layered silicate nanocomposite as high performance ablative materials. *Applied Clay Science*, **15** (1999), 67–92.

[28] J. H. Koo, H. Stretz, A. Bray, J. Weisphennig, Z. P. Luo, and R. Blanski. Nanostructured materials for rocket propulsion system: recent progress. In

Proceedings of the 44th AIAA/ASME/ASCE/AHS structures, structure Dynamics, and Materials Conference, April 7–10, 2003, Norfolk, Virginia, Paper AIAA 2003–1769.

[29] G. Beall, Z. Shirin, S. Harris, M. Wooten, C. Smith, and A. Bray. Development of an ablative insulation material for ramjet applications. *Journal of Spacecraft and Rockets*, **41** (2004), 1068–1071.

[30] Y. Xu, J. Yang, and Y. Xu. Thermooxidation effects on the structuring of molten polypropylene–clay nanocomposites. *Polymer International*, **55** (2006), 681–687.

[31] S.-J. Park, K. Li, and S.-K. Hong. Preparation and characterization of layered silicate-modified ultrahigh–molecular-weight polyethylene nanocomposites. *Journal of Industrial and Engineering Chemistry*, **11** (2005), 561–566.

[32] H. E. Kissinger. Reaction kinetics in differential thermal analysis. *Analytical Chemistry*, **29** (1957), 1702.

[33] R. Zong, Y. Hu, N. Liu, S. Wang, and G. Liao. Evaluation of the thermal degradation of PC/ABS/montmorillonite nanocomposites. *Polymer Advanced Technology*, **16** (2005), 725–731.

[34] E. S. Freeman and B. Carroll. The application of thermoanalytical techniques to reaction kinetics. The thermogravimetric evaluation of the kinetics of the decomposition of calcium oxalate monohydrate. *Journal of Physical Chemistry*, **62** (1958), 394.

[35] H. Qin, S. Zhang, C. Zhao, and M. Yang. Zero-order kinetics of the thermal degradation of polypropylene/clay nanocomposites. *Journal of Polymer Science: Part B: Polymer Physics*, **43** (2005), 3713–3719.

[36] A. W. Coats and J. P. Redfern. Kinetic parameters from thermogravimetric data. *Nature*, **201** (1964), 68–69.

[37] D. K. Chattopadhyay, A. K. Mishra, B. Sreedhar, and K. V. S. N. Raju. Thermal and viscoelastic properties of polyurethane-imide/clay hydbrid coatings. *Polymer Degradation and Stability*, **91** (2006), 1837–1849.

[38] C. A. Wilkie and P. Jash. Effects of surfactants on the thermal and fire properties of poly(methyl methacrylate)/clay nanocomposites. *Polymer Degradation and Stability*, **88** (2005), 401–406.

[39] M. Modesti, D. Lorenzetti, and B. S. Besco. Thermal behavior of compatibilised polypropylene nanocomposite: effect of processing conditions. *Polymer Degradation and Stability*, **91** (2006), 672–680.

[40] S. Su, D. D. Jiang, and C. A. Wilkie. Study on the thermal stability of polystyryl surfactants and their modified clay nanocomposites. *Polymer Degradation and Stability*, **84** (2004), 269–277.

[41] F. Gong, M. Feng, C. Zhao, S. Zhang, and M. Yang. Thermal properties of poly (vinyl chloride)/montmorillonite nanocomposites. *Polymer Degradation and Stability*, **84** (2004), 289–294.

[42] S. Yoshimoto, F. Ohashi, and T. Kameyama. Characterization and thermal degradation studies on polyaniline–intercalated montmorillonite nanocomposites prepared by a solvent-free mechanochemical route. *Journal of Polymer Science: Part B: Polymer Physics*, **43** (2005), 2705–2714.

[43] S. K. Srivastava and P. H. Acharya. Ethylene/vinyl acetate copolymer/clay nanocomposites. *Journal of Polymer Science: Part B: Polymer Physics*, **44** (2006), 471–480.

[44] V. Babrauskas. Development of the cone calorimeter A bench-scale heat release rate apparatus based on oxygen consumption. (NBSIR 82–2611). (US) National Bureau of Standards, 1982.

[45] V. Babrauskas and G. Mulholland. Smoke and soot data determinations in the cone calorimeter. In Mathematical Modeling of Fires (ASTM STP 983). *American Society for Testing and Materials*, Philadelphia (1987), 83–104.

[46] The cone calorimeter test. ISO 5660–1 1993.

[47] A. B. Morgan. Flame retarded polymer layered silicate nanocomposites: a review of commercial and open literature systems. *Polymer Advanced Technology*, **17** (2006), 206–217.

[48] J. W. Gilman, R. H. Harris Jr, J. R. Shields, T. Kashiwagi, and A. B. Morgan. A study of the flammability reduction mechanism of polystyrene-layered silicate nanocomposite: layered silicate reinforced carbonaceous char. *Polymer Advanced Technology*, **17** (2006), 263–271.

[49] J. W. Gilman, T. Kashiwagi, A. B. Morgan, R. H. Harris, L. Brassell, and M. J. VanLandingham. 2005. NISTIR 6531. (National Institute of Standards and Technology, Washington DC, 2005).

[50] M. Lewin, E. M. Pearce, K. Levon, A. Mey-Marom, M. Zammarano, C. A. Wilkie, and B. N. Jang. Nanocomposites at elevated temperatures: migration and structural changes. *Polymer Advanced Technology*, **17** (2006), 226–234.

[51] S. Su, D. D. Jiang, and C. A. Wilkie. Methacrylate modified clays and their polystyrene and poly(methyl methacrylate) nanocomposites. *Polymer Advanced Technology*, **15** (2004), 225–231.

[52] X. Zheng, D. D. Jiang, D. Wang, and C. A. Wilkie. Flammability of styrenic polymer clay nanocomposites based on a methyl methacrylate oligomerically-modified clay. *Polymer Degradation and Stability*, **91** (2006), 289–297.

[53] J. Zhang and H. P. Zhang. Study on the flammability of HIPS-montmorillonite nanocomposites prepared by static melt intercalation. *Journal of Fire Sciences*, **23** (2005), 193–208.

[54] S. Bourbigot, D. L. Vanderhart, J. W. Gilman, S. Bellayer, H. Stretz, and D. R. Paul. Solid state NMR characterization and flammability of styrene–acrylonitrile copolymer montmorillonite nanocomposite. *Polymer*, **45** (2004), 7627–7638.

[55] G. Chigwada, D. Wang, and C. A. Wilkie. Polystyrene nanocomposites based on quinolinium and pyridinium surfactants. *Polymer Degradation and Stability*, **91** (2006), 848–855.

[56] G. Chigwada, D. Wang, D. D. Jiang, and C. A. Wilkie. Styrenic nanocomposites prepared using a novel biphenyl-containing modified clay. *Polymer Degradation and Stability*, **91** (2006), 755–762.

[57] S. Su, D. D. Jiang, and C. A. Wilkie. Polybutadiene-modified clay and its nanocomposites. *Polymer Degradation and Stability*, **84** (2004), 279–288.

[58] G. Chigwada, D. D. Jiang, and C. A. Wilkie. Polystyrene nanocomposites based on carbazole-containing surfactants. *Thermochimica Acta*, **436** (2005), 113–121.

[59] G. Beyer. Flame retardancy of nanocomposites based on organoclays and carbon nanotubes with aluminum trihydrate. *Polymer Advanced Technology*, **17** (2006), 218–225.

[60] B. Schartel, U. Knoll, A. Hartwig, and D. Putz. Phosphonium-modified layered silicate epoxy resins nanocomposites and their combinations with ATH and organo-phosphorus fire retardants. *Polymer Advanced Techology*, **17** (2006), 281–293.

[61] N. Ristolaninen, U. Hippi, J. Seppala, A. Nykanen, and J. Ruokolainen. Properties of polypropylene/aluminum trihydroxide composites containing nanosized organoclay. *Polymer Engineering Science*, **45** (2005), 1568–1575.

[62] H. Lu, Y. Hu, J. Xiao, Z. Wang, Z. Chen, and W. Fan. Magnesium hydroxide sulfate whisker flame retardant polyethylene/montmorillonite nanocomposites. *Journal of Materials Science*, **41** (2006), 363–367.

[63] H. Zhang, Y. Wang, Y. Wu, L. Zhang, and J. Yang. Study on flammability of montmorillonite/styrene-butadiene rubber (SBR) nanocomposites. *Journal of Applied Polymer Science*, **97** (2005), 844–849.

[64] Y. Tang, Y. Hu, B. Li, L. Liu, Z. Wang, Z. Chen, and W. Fan. Polypropylene/ montmorillonite nanocomposites and intumescent, flame-retardant montmorillonite synergism in polypropylene nanocomposites. *Journal of Polymer Science: Part A: Polymer Chemistry*, **42** (2004), 6163–6173.

[65] Z.-L. Ma, W.-Y. Zhang, and Z.-Y. Liu. Using PA6 as a charring agent in intumescent polypropylene formulations based on carboxylated polypropylene compatibilizer and nano-montmorillonite synergistic agent. *Journal of Applied Polymer Science*, **101** (2006), 739–746.

[66] Y. Tang, Y. Hu, L. Song, R. Zong, Z. Gui, and W. Fan. Preparation and combustion properties of flame retarded polypropylene–polyamide-6 alloys. *Polymer Degradation and Stability*, **91** (2006), 234–241.

[67] S. Nazare, B. K. Kandola, and A. R. Horrocks. Flame-retardant unsaturated polyester resin incorporating nanoclays. *Polymer Advanced Technology*, **17** (2006), 294–303.

[68] Y. Hu, Y. Tang, and L. Song. Poly(propylene)/clay nanocomposites and their application in flame retardancy. *Polymers for Advanced Technologies*, **17** (2006), 235–245.

[69] H. Ma, Z. Fang, and L. Tong. Preferential melt intercalation of clay in ABS/ brominated epoxy resin-antimony oxide (BER-AO) nanocomposites and its synergistic effect on thermal degradation and combustion behavior. *Polymer Degradation and Stability*, **91** (2006), 1972–1979.

[70] G. Chigwada, P. Jash, D. D. Jiang, and C. A. Wilkie. Synergy between nanocomposite formation and low levels of bromine on fire retardancy in polystyrene. *Polymer Degradation and Stability*, **88** (2005), 382–393.

[71] D. Wang, K. Echols, and C. A. Wilkie. Cone calorimetric and thermogravimetric analysis evaluation of halogen-containing polymer nanocomposites. *Fire and Materials*, **29** (2005), 283–294.

[72] L. Song, Y. Hu, Z. Lin, S. Xuan, S. Wang, Z. Chen, and W. Fan. Preparation and properties of halogen-free flame-retarded polyamide 6/organoclay nanocomposite. *Polymer Degradation and Stability*, **86** (2004), 535–540.

Index